IntechOpen

*Supporting open minds since 2005*

Dynamic Data Assimilation - Beating the Uncertainties
http://dx.doi.org/10.5772/intechopen.87789
Edited by Dinesh G. Harkut

Contributors
Yassine Zahraoui, Mohamed Akherraz, Afef Salhi, Ahmed Fakhfakh, Fahmi Ghozzi, Anand Raju, Shanthi Thirunavukkarasu, Wesam Salah Alaloul, Hong Son Hoang, Dinesh G. Harkut, Remy Baraille, Abdul Hannan Qureshi



First published in London, United Kingdom, 2020 by IntechOpen
IntechOpen is the global imprint of INTECHOPEN LIMITED, registered in England and Wales, registration number: 11086078, 5 Princes Gate Court, London, SW7 2QJ, United Kingdom
Printed in Croatia

British Library Cataloguing-in-Publication Data
A catalogue record for this book is available from the British Library

Additional hard and PDF copies can be obtained from orders@intechopen.com

Dynamic Data Assimilation - Beating the Uncertainties
Edited by Dinesh G. Harkut
p. cm.
Print ISBN 978-1-83968-083-0
Online ISBN 978-1-83968-084-7
eBook (PDF) ISBN 978-1-83968-085-4

For EU product safety concerns:
IN TECH d.o.o., Prolaz Marije Krucifikse Kozulić 3, 51000 Rijeka, Croatia, info@intechopen.com or visit our website at intechopen.com.

# Meet the editor

Dr. Dinesh G. Harkut is Associate Professor at Prof Ram Meghe College of Engineering & Management (PRMCEAM), Badnera, India, in the Computer Science and Engineering Department. He obtained a bachelor's degree, a master's of engineering (CSE), and a PhD (CSE) from SGBAU Amravati University, Maharashtra, India. He also holds a master's degree and PhD in Business Administration. His primary research interests are in artificial intelligence, big data, analytics, embedded systems, and e-commerce. He has supervised eighteen master's degree and twenty-four bachelor's degree students. He has published forty-seven papers in refereed journals and published six books with international publishers. He has also organized various workshops, sessions, conferences, and trainings. He has two patents filed and published in his name in India. He is a member of the Board of Studies (Computer Science and Engineering) and a recognized PhD supervisor at SGBAU Amravati University, Maharashtra, India. He holds membership in various professional bodies including the Institution of Electronics and Telecommunication Engineers (IETE), New Delhi; International Society for Technology in Education (ISTE), New Delhi; Universal Association of Computer and Electronics Engineers (UACEE), USA; International Economics Development and Research Center (IEDRC), Hong Kong; International Association of Engineers (IAENG), Hong Kong; and the European Alliance for Innovation, Belgium.

# Contents

# Preface

*It's alright, sweetie.*
*In the information age of ridiculously enormous and complex data set,*
*everybody feels stupid, unless one has the right tools and methodology to deal with.*
*Anonymous*

This book provides a comprehensive introduction to data assimilation, a vital tool used mostly in atmospheric science and oceanography. Ensemble data assimilation methods have been applied with remarkable success in several real-life history-matching problems. However, performance is severely degraded as data assimilation methods are based on Gaussian assumptions. This problem can be overcome with artificial neural networks, machine learning, and deep learning. The synergy of these complementary technologies leverages their benefits and results in the emergence of one of the most efficient tools for handling linear and non-linear models predicting the evolution of the atmosphere. This hybrid approach emulates hidden, chaotic dynamics and predicts future states with desired accuracy. The most known use of data assimilation is predicting the state of the atmosphere using meteorological data. Data assimilation is a vital step in numerical modeling, specifically in the atmospheric sciences and oceanography. However, even with a good understanding of the underlying physical laws that drive it, its chaotic nature makes it extremely difficult to determine the state of the environment, specifically atmospheric variables like temperature, humidity, pressure, and so on, with accuracy in a given spatio-temporal domain. This book presents the material in a clear, simple style and examines the many challenges and opportunities in the field of data assimilation.

I would like to convey our appreciation to all authors for their contributions. I owe special thanks to Author Service Managers Ms. Ivana Barac and Ms. Sara Debeuc, and Commissioning Editor Ms. Klara Mestrovic, at IntechOpen, London, UK, for their kind support and great efforts in bringing this book to fruition. In addition, I am grateful to all those who worked behind the scenes and assisted in formatting the book.

**Dr. Dinesh G. Harkut**
Dean and Associate Professor,
Department of Computer Science and Engineering,
Prof Ram Mehge College of Engineering and Management,
Badnera-Amravati, M.S., India

Chapter 1

# Introductory Chapter: Data Assimilation

*Dinesh G. Harkut*

## 1. Introduction

Our life is highly influenced and affected by the uncertainty in predicting the outcome of various phenomena and human activities. All the activities are highly influenced by predictions like uncertainty in predicting natural phenomena like rains, heat waves, short-term climate change, cyclone, tornados, or revenue prediction/projection by state/central government while preparing budgets and, GDP growth while formulating financial policies or predicting stock prices/indices by individual investors. These predications are based on some relevant class of models that are either causality or empirically derived and can be:

1. Static model or dynamic model

2. Stochastic model or deterministic model

3. Based on either continuous space or discrete space

4. Operates in either discrete time or continuous time domain

Irrespective of the model used or its origin, solution computed or predictions generated were based on several prerequisite unknown controlling parameters along with initial conditions, boundary conditions variables those are based on some estimation: observations of the phenomenon in question like observed pressure distribution around the eye of the hurricane, data from radars or satellites, the time series of data on unemployment, etc.

Data assimilation is basically a process of fusing data with the model for the singular purpose of estimating the unknown variables. One can obtain an instantiation of the model once these estimates are available, which in turn then run forward in time to generate the requisite forecast products for public consumption. Basic mathematical principles and tools along with conventional methods like Kalman filters and variational approaches, which find applications in dynamic data assimilation include: linear algebra, multivariate calculus, estimation theory, finite dimensional optimization theory, chaos, and nonlinear dynamics. It refers to the computation of the conditional probability distribution function of the output of a numerical model describing a dynamical process, conditioned by observations. Numerical prediction of atmospheric evolution is critically dependent on the initial conditions provided to it. It is a technique by which numerical model data and observations are combined to obtain an analysis that best represents the state of the phenomena of interest. It is the process of updating model forecasts (priors) with information



IntechOpen

from the observations of complete or incomplete state variables, that is, combining a physical model with observations with a goal to produce an improved model state (posteriors), which better represents the dynamics system.

Data assimilation can be used for multiple purposes: to estimate the optimal state of a model and to estimate the initial state of a system in order to use it to predict the future state of the system. The most known use of data assimilation is predicting the state of the atmosphere using meteorological data. Data assimilation is a vital step in numerical modeling, specifically in the atmospheric sciences and oceanography. However, even with a good understanding of the underlying physical laws that drive it, the chaotic nature makes it extremely difficult to determine the state of environment specifically all the atmospheric variables such as temperature, humidity, and pressure with desire resolutions and accuracy in given spatio-temporal domain. Data assimilation is a crucial step in numerical modeling, particularly in the oceanography and atmospheric sciences, which involves huge volume and variety of observations of either complete or incomplete state variables conditions. The volume and variety involved further increase the complexity of the task. Conventional methods for assimilation include Kalman filters and variational approaches which addresses redundancy and uneven spatial or temporal distribution of data which in turn can consume massive data sets. However, because these methods rely on Gaussian assumptions, performance is severely degraded when the prior facts are described in terms of complex distributions and based on unrealistic assumptions, particularly linearity and normality. Nevertheless, these approaches are incapable of overcoming fully their unrealistic assumptions, particularly linearity, normality, Markovian processes, knowledge of underlying mathematical models, and zero error covariances. Predicting the evolution of the atmosphere is a complicated problem that requires the most accurate initial conditions to obtain an accurate estimate of the atmospheric state variables at a given time and point. Though lots of information through meteorological observations from various sources like weather stations, radio soundings, and ocean buoys is easily available, but it is not enough to fully describe the conditions of the model and also the observations may contain errors. The data from sensors are often partial, distorted, or too inaccurate. State observations of the atmosphere can be corrected to some extend by taking samples in different time and space. Linking actual sensors data with physical model of the atmosphere facilitates debugging the errors by correcting the initial conditions and thus finding the missing part of model dynamics.

Furthermore, ensemble data assimilation method gives significant results in most of the real-life history/data-matching problem domain. Kalman filtering has been a robust method for the past few decades which is further complemented by the recent advances in such filters specifically the use of ensembles and the extended Kalman filter. It combines observation data and the underlying dynamical principles governing the system to provide an estimate of the state of the system which is better than could be obtained using just the data or the model alone. But, despite of popularity, Kalman filters and ensemble Kalman filters are suboptimal as it is based on some unrealistic assumptions like correctness about the prior knowledge and the number of ensemble members, linearity, error covariances and are inefficient when the data sets become large.

Though traditional data assimilation methods introduce Kalman filters and variational approaches, application of artificial intelligence, neural network, machine learning, and cognitive computing can be exploited further to forecast by accommodating the dynamics of model to obtain the most critical initial condition precisely. Recent progress in machine learning has shown how to forecast and, to some extent, learn the dynamics of a model from its output, resorting in particular to neural networks and deep learning techniques. Specifically, the use of machine learning combine data with human knowledge in the form of mechanistic models facilitates forecasting future states, to attribute missing data from the past by smoothing and to infer measurable and unmeasurable quantities with a desired accuracy. In the last decades, the volume and quality of observations from land, ocean atmosphere, and space-based platform lead to massive amounts of data available to incorporate into models have increased dramatically, particularly thanks to remote sensing. At the same time, new developments in machine learning, particularly deep learning, have demonstrated impressive skills in reproducing complex spatiotemporal processes by efficiently using a huge amount of data, thus paving the path for their use in Earth System Science.

The accuracy, efficiency, speed, and scalability in recovering state trajectories ascertain the feasibility of machine learning for data assimilation. This complementary combination and arrangements of the two technologies will enhance the sophistication to justify their application requirements, thwart their implementation issues and improve the accuracy.

Machine learning may complimentary and provide efficient alternative to Kalman filtering to predict the future of a dynamic system without any knowledge of the underlying physical model and make minimal assumptions about the data and error properties. Data assimilation and machine learning are complimentary to each other and deep learning which makes it possible to predict and understand complex spatio-temporal phenomena, sometimes in an optimal way, compared to traditional data assimilation approaches. This combination of two techniques enables us to obtain much better results by exploiting the purely physical approach of data assimilation which is most suitable for linear parts of model and purely given approach of machine learning which facilitates the observations and address the nonlinear parts of the model. Effectiveness of machine learning trained on noise-free and complete observations of the system in reconstructing the model dynamics have been proven by various numerical models and hence incorporating explicit or implicit regularization processes, machine learning algorithms aids in optimizing in high-dimension without the need for additional information under the form of an explicit prior. Machine learning even finds its application in the situations where either the observations are subsampled in time or only a dense portion of the system is observed or when. Thus, to leverage the benefits of recent machine learning developments, which in turn provide flexibility and facilitate parallel calculations, a novel hybrid method the combination of data assimilation and machine learning finds wide spread application in recent time. This hybrid model has dual edge benefits of predicting future state by emulating hidden chaotic dynamics.

Thus, the use of enhanced and cheaper computational capability and the successful synergy between data assimilation and machine learning, two seemingly unrelated inverse problems have proven here with a low-dimensional system, encourages further investigation of such hybrids with more sophisticated dynamics and proven with a low-dimensional system.

## Author details

Dinesh G. Harkut
Prof Ram Meghe College of Engineering and Management, Badnera–Amravati (M.S.), India

*Address all correspondence to: dg.harkut@gmail.com

Chapter 2

# Adaptive Filter as Efficient Tool for Data Assimilation under Uncertainties

*Hong Son Hoang and Remy Baraille*

## Abstract

In this contribution, the problem of data assimilation as state estimation for dynamical systems under uncertainties is addressed. This emphasize is put on high-dimensional systems context. Major difficulties in the design of data assimilation algorithms is a concern for computational resources (computational power and memory) and uncertainties (system parameters, statistics of model, and observational errors). The idea of the adaptive filter will be given in detail to see how it is possible to overcome uncertainties as well as to explain the main principle and tools for implementation of the adaptive filter for complex dynamical systems. Simple numerical examples are given to illustrate the principal differences of the AF with the Kalman filter and other methods. The simulation results are presented to compare the performance of the adaptive filter with the Kalman filter.

**Keywords:** adaptive filter, innovation process, minimum prediction error, simultaneous perturbation stochastic approximation, filter stability

## 1. Introduction

In this chapter, the adaptive filter (AF) is considered as a computational device that yields estimates of the system state by minimizing recursively (in time) the error between the predicted output of the device and its observed signal in real time. As the main objective of the AF is to produce estimates of the state in high-dimensional systems (HdSs), we shall focus the attention on the mathematical form of the AF in a state-space form as that used in the Kalman filter (KF) [1]. In this chapter, the HdS is referred to as a system whose state dimension is of order $O(10^7) - O(10^8)$.

The assimilation problem in this chapter is formulated as a standard filtering problem. For simplicity, let the dynamical system be described by the equation

$$x(t+1) = \Phi x(t) + w(t), x(0) = x_0, t = 0, 1, \ldots \quad (1)$$

where $x(t)$ is the system state at the $t$ time instant. At each time instant $t$, we are given the observation for the system output

$$z(t+1) = Hx(t+1) + v(t+1), t = 0, 1, \ldots \quad (2)$$

In (1) and (2), $w(t)$ is the model error (ME), $v(t)$ is the observation error (ObE), and $\Phi$ represents the system dynamics. In general, the system (1) and (2) may be nonlinear with $\Phi x = f(x)$, $Hx = h(x)$. The filtering problem for a partial observed dynamical system (1) and (2) is to obtain the best possible estimate for the state $x(t)$ at each instant $t$, given the set of observations $Z(1:t) = [z(1), \dots, z(t)]$.

There exist different techniques to solve estimation problems. The simplest approach is related to linear estimator [2], since it requires only first two moments. Linear estimation is frequently used in practice when there is a limitation in computational complexity. Among others, the widely used methods are maximum likelihood, least squares, method of moments, the Bayesian estimation, minimum mean square error (MMSE), etc. For more details, see [3].

There are limitations of optimal filters. In practice, the difficulties are numerous: the statistics of signals which may not be available or cannot be accurately estimated; there may not be available time for statistical estimation (real-time); the signals and systems may be non-stationary; memory required and computational load may be prohibitive. All these difficulties become insurmountable, especially for HdSs.

In order to deal with real-time applications, the AFs appear to be a valuable tool in solving estimation problems when there is no time for statistical estimation and when we are dealing with non-stationary signals and/or systems environment. They can operate satisfactorily in unknown and possibly time-varying environments without user intervention. They improve their performance during operation by learning statistics from current signal observations. Finally, they can track variations in the signal operating environment [4].

It is well-known that the MMSE estimator in the class of Borel measurable (with respect to (wrt) $Z(1:t)$) functions is given by the conditional mean

$$\hat{x}(t) = E[x(t)/Z(1:t)] \tag{3}$$

Under standard conditions, related to the noise sequences $w(t), v(t)$ (Gaussian i.i.d.—identically independent (temporal) distributed), the estimate (3) $\hat{x}(t)$ for $x(t)$ can be obtained from the KF in the recursive form

$$\hat{x}(t+1) = \hat{x}(t+1/t) + K\zeta(t+1), \tag{4}$$

$$\hat{x}(t+1/t) = \Phi\hat{x}(t), \zeta(t+1) = z(t+1) - \hat{x}(t+1/t) \tag{5}$$

$$K(t) = M(t)H^T\left[HM(t)H^T + R\right]^{-1} \tag{6}$$

$$M(t+1) = \Phi P(t)\Phi^T + Q \tag{7}$$

$$P(t) = [I - K(t)H]M(t)[I - K(t)H]^{\mathrm{T}} + K(t)RK^T(t) \tag{8}$$

In (4)–(8), $Q, R$ are the covariance matrix for $w(t)$ and $v(t)$, respectively. One sees that $K = K(M)$—the gain matrix, is a function of $M := M(t+1)$—the error covariance matrix (ECM) for the state prediction error (PE) $e(t+1/t)$ which is defined as $e(t+1/t) := x(t+1) - \hat{x}(t+1)$, and $\zeta(t+1)$ is known as innovation vector. Note from (7) and (8) that the ECM $M(t+1)$ can be found by solving the matrix nonlinear Algebraic Riccati equation (ARE). Generally speaking, a solution of the ARE is not unique. Conditions must be introduced for ensuring an existence of a unique non-negative definite solution [5]. It is remarkable that the ECMs $P, M$ in (7) and (8) do not depend on observations; therefore, they can be computed in advance, offline, given the system matrices and the noise covariances. The same remark is valid for the gain matrix $K$ in (6). In contrast, the gain in the AF is observation-dependent [6] (see Section 2).

Under the most favorable conditions (perfect knowledge of all system parameters and noise statistics), for a dynamical system with dimension of order $10^7$–$10^8$, it is impossible to solve the ARE (due to computational burden), not to say about storing $M, P$. To overcome these difficulties, the AF is proposed. Mention that the KF is also an MMSE filter in the complete Hilbert space of random variables, $E|\xi|^2 < \infty$, with scalar product $(\xi_1, \xi_2)_H = E < \xi_1 \xi_2 >$ and the norm $||\xi|| = \left[E\left(|\xi|^2\right)\right]^{\{1/2\}}$.

For the nonlinear models, there are KF variants, among those are the extended KF (EKF) [7], the unscented KF (UKF, [8]), and the Ensemble Kalman filter (EnKF, [9]). In the EnKF, the ECM is a sampled ECM whose samples are generated using samples of the state variable, and consequently the ECM in the KF becomes a sampled ECM. For an example of application of the EnKF for data assimilation in geophysical data assimilation with high dimensional model, see [10]. Another class of ensemble filtering technique is a class of particle filters (PF, [11]). The basic idea of the PF (also the EnKF) is to use a discrete set of weighted $n$ particles to represent the distribution of $x(t)$, where the distribution is updated at each time by changing the particle weights according to their likelihoods.

Despite a possible implementation of the KF variants, they might still be seriously biased because the accuracy of the KF update requires linearity of the observation function and Gaussianity of the distribution of system state $x(t)$. In reality, the KF (4)–(8) may be biased and unstable, even divergent [12]. Today, the PF algorithms are ineffective for HdS data assimilation.

In this chapter, we shall show how the AF can be efficient in dealing with uncertainties existing in the filtering problem (1) and (2). In Section 2, a brief outline of the AF is given. The main features of the AF, which are different to those of the KF, are presented. This concerns the optimal criteria, stabilizing gain structure, optimization algorithms. Section 3 shows in detail how the AF is capable of dealing with uncertainties in the specification of ME covariance. The hypothesis on a subspace of ME is presented in Section 4 from which one sees how one can make order reduction for representing the bias and ME covariance. Simple numerical examples on one- and two-dimensional systems are given in Section 5 to illustrate in details the differences between the AF and the traditional KF. Numerical experiments on low and high dimensional systems are given in Section 6 to demonstrate how the AF algorithm works. The performance comparison between the AF and KF, for both situations of perfect knowledge of ME statistics and that with ME uncertainties, is also presented. Conclusions and perspectives of the AF are summarized in Section 7.

## 2. Adaptive filter

The AF is originated from [13]. It is constructed for estimating the state of a dynamical system based on partially observed quantities related in some way to the system state. As reported before, for linear systems contaminated by Gaussian noise, the MMSE estimate can be obtained by the KF. Since publication of [1] in 1960, an uncountable number of works are done for solving engineering problems by KF, in all engineering fields, as well as many modifications have been proposed. The reasons for the need in modification of the KF are numerous, but mostly related to nonlinear dynamics, parameter uncertainties in specification of system parameters, bias of ME, unknown statistics of ME, model reduction. With the rapid progress of computer technology (computer power, memory, ... ), various simplified versions of KF are suggested for solving filtering (or data assimilation) for HdSs, in particular, in meteorology and oceanography.

Direct application of the KF to HdSs is impossible due to the limit in computer power, memory, and computational time. In particular, the KF requires to solve the matrix AREs (7) and (8) for computing ECMs $M(t), P(t)$. Storing such matrices is impossible, not to say on computational time.

Different simplified approaches are proposed for overcoming difficulties in the application of the KF. The example of successful tool for solving data assimilation problems in HdSs is the EnKF [9]. In the EnKF, an ensemble of error samples, of small size, is generated on the basis of model states to approximate the ECMs. In practice of data assimilation for HdSs, it is possible to generate only ensembles of moderate sizes (of order $O(100)$) by model integrations over the assimilation window (time interval between two successive arrivals of observations) since one such integration takes several hours! The other approach like PF is based on sampling from conditional distributions. Theoretically, this approach is more appropriate for nonlinear problems because no linearization is required as in the EKF (Extended KF based on linearization technique). However, even for filtering problems with state dimensions of order $O(10)$, relatively large ensembles of size $O(10000)$ would be required in order to produce reasonably good performance.

The AF in [13] is based on the different idea. Here, no linearization is required for nonlinear filtering problems. For the problem (1) and (2), the filter is of the form (4) and (5) but the gain $K = K(\theta)$ is assumed to be of a given stabilizing structure [6]. It means that $K$ is parametrized by some vector of unknown parameters $\theta \in \Theta$ so that the filter (4) and (5) with the gain $K(\theta), \forall \theta \in \Theta$ is stable. It is well-known that under mild conditions, the solution of the ARE will tend (quickly) to stationary solution $M_\infty$ and so the gain (6), to the stationary gain $K_\infty$. Moreover, the innovation $\zeta(t+1) = z(t+1) - \hat{z}(t+1/t)$, representing the error for the output prediction $\hat{z}(t+1/t) := H\hat{x}(t+1/t) = H\Phi\hat{x}(t)$, is unbiased and of minimum variance. This fact leads to the idea to seek the optimal vector $\theta$ by minimizing

$$J(\theta) := E[\Psi(\zeta)] \to \arg\ \min \theta \tag{9}$$

here $E[.]$ denotes the *average* in a *probabilistic sense*. For stationary systems (1) and (2), if we assume the validity of the ergodic hypothesis, the *average* value in a *probabilistic sense*, expressed in (7), is almost everywhere equivalent to the time average (for large time of running the dynamical system). The optimal $\theta^*$ can be found by solving the equation

$$\nabla_\theta J(\theta) = \nabla_\theta E[\Psi(\zeta)] = E[\nabla_\theta \Psi(\zeta)] = 0 \tag{10}$$

A stochastic approximation (SA) algorithm for solving (10) can be written out

$$\theta(t+1) = \theta(t) - \gamma(t)\nabla_{\theta(t)}\Psi[\zeta(t+1)] \tag{11}$$

Conditions related to the sequence of positive scalar $\gamma(t)$ for ensuring a convergence $\{\theta(t)\}$ in the procedure (11) are

$$\gamma(t) > 0, \sum_{t=0}^{\infty} \gamma(t) = \infty, \sum_{t=0}^{\infty} \gamma^2(t) < \infty \tag{12}$$

One of the most advantages of the SA algorithm (11) is that, instead of computing the gradient of the cost function (9) (which requires knowledge of probability distribution), the algorithm (11) is based on the knowledge of only the gradient of sample cost function $\Psi$ (wrt to $\theta$) which can be easily evaluated numerically.

*Comment 2.1.* Generally speaking, the convergence rate of the algorithm (11) and (12) is $O(1/t)$. It is possible to improve the convergence rate of the SA by averaging of the iterates,

$$\theta_m(t) = \frac{1}{t}\sum_{t'=1}^{t} \theta(t') \tag{13}$$

For more details, see [14].

*Comment 2.2.* For high HdSs, even with $\theta$ being of moderate dimension, instead of the algorithm (12) or (13), the SPSA (Simultaneous Perturbation Stochastic Perturbation) algorithms in [15, 16] are of preference. That is due to the fact that integration of HdS over the assimilation window is very expensive. These algorithms generate stochastic perturbation $\delta\theta = (\delta\theta_1, \ldots, \delta\theta_n)^T$ with components as Bernoulli i.i.d. realizations. Each $i^{th}$ component of the gradient-like (pseudo-gradient) vector is computed as the divided difference $\delta\Psi/\delta\theta_i$ where $\Psi := \Psi[\theta + \delta\theta] - \Psi[\theta]$. This allows to evaluate the gradient-like vector by only two or three time integration of the numerical model.

For details on the SPSA algorithm and its convergence rate, see [15, 16].

## 3. Covariance uncertainties and AF

### 3.1 Covariance uncertainties

#### *3.1.1 Adjoint approach*

As seen from (11) and (12), implementation of SA algorithms is much simpler for searching optimal gain parameters compared to the other optimization methods. The SA algorithms require only numerical computing derivatives $\nabla_\theta \Psi[\zeta(t+1)]$ of the sample cost function $\Psi[\zeta(t+1)]$ wrt $\theta$ evaluated at $\theta(t)$ and $\gamma(t)$ is a scalar which can be chosen a priori, for example, as $\gamma(t) = \frac{1}{t}$. That is possible due to introducing the ergodic hypothesis on of the system (1) and (2) from which there exists an asymptotic optimal gain

First, consider the situation when the vector of parameters consists of are all elements of $K, \theta = K$. Compute the innovation vector,

$$\zeta(t+1) = z(t+1) - \hat{z}(t+1/t) = z(t+1) - H\Phi\hat{x}(t)$$

$$\hat{x}(t) = \hat{x}(t-1/t) + K(\theta)\zeta(t)$$

$$\delta_\theta\zeta(t+1) = -H\Phi\delta_\theta\hat{x}(t) = -H\Phi\delta_\theta K(\theta)\zeta(t) = -H\Phi\delta K\,\zeta(t).$$

$$\delta_K\Psi[\zeta(t+1)] = \delta_K < \zeta(t+1), \zeta(t+1) > \; = 2 < \zeta(t+1), \delta_K\zeta(t+1) >$$

$$= -2 < \zeta(t+1), H\Phi\delta K\zeta(t) > \; = -2 < \Phi^T H^T\zeta(t+1), \delta K\zeta(t) > .$$

Let us compute derivatives of the sample cost function $\Psi$ wrt the elements $K_{ij}$ of the gain $K$. To do so, one needs to integrate the adjoint operator $\Phi^T$ s.t. the forcing $f := -H^T\zeta(t+1)$ which yields $\psi := -\Phi^T H^T\zeta(t+1)$ and hence

$$\delta\Psi/\delta K_{ij} = -\psi_i\zeta(t,j), \tag{14}$$

here $\psi_i$ is the $i^{th}$ component of $\psi$, $\zeta(t,j)$ the $j^{th}$ component of $\zeta(t)$. The AF now takes the form

$$\hat{x}(t+1) = \hat{x}(t+1/t) + K\zeta(t+1), \quad (15)$$

$$\hat{x}(t+1/t) = \Phi\hat{x}(t), \zeta(t+1) = z(t+1) - H\hat{x}(t+1/t), \quad (16)$$

$$K(t+1) = K(t) - \gamma(t)\nabla_K\Psi[\zeta(t+1)], \quad (17)$$

where $\nabla_K\Psi(\zeta(t+1))$ is the gradient vector whose components are computed by (14). In the AF (15)–(17), no matrix ARE (see (7) and (8) in the KF) is involved. The AF (15)–(17) is quite realizable for HdSs, since at each assimilation instant we need to integrate only the direct model to produce the forecast (16) and (eventually) an adjoint model over the assimilation window for computing $\nabla_K\Psi[\zeta(t+1)]$ [13].

### *3.1.2 Simultaneous perturbation stochastic approximation (SPSA) approach*

Remark that in the form (14) the adjoint operator $\Phi^T$ would be available to implement the AF. It is well-known that construction of numerical code for $\Phi^T$ is a very difficult and heavy task, especially for meteorological and oceanic numerical models which are HdSs and nonlinear (linearization is required).

A comparison study of the AF with other assimilation methods is done in [17]. Compared to the AF, the widely used variational method (VM) minimizes the distance between the observations available (for example, the observations of the whole set $Z[1:T]$) and the outputs of the dynamical system. This optimization problem is carried out in the phase space, hence is very difficult and expensive. Theoretically, a simplification is possible subject to (s.t.) the condition of linearity of the dynamical system: in this case, one can reformulate the VM minimization problem as searching the best estimate for the initial system state $x(0)$. For HdSs, to ensure a merely high quality estimate for $x(0)$, it is necessary: (i) to take the observation window as large as possible; (ii) to parameterize the initial state by some parameters (using a slow manifold, for example). Iterative minimization procedures require usually $O(10)$ iterates involving integrating the direct and adjoint models over the window $[1, T]$. For an unstable dynamics, integration of direct and adjoint equations over a long period naturally amplify the initial errors during assimilation process. For a more detailed comparison between the AF and VM, see [17].

Thus, if the ergodic conditions hold, there exists an optimal stationary gain and the AF in limit will approach to the optimal one *in the given class of stable filters*. It is important to emphasize that up to this point, no covariance matrices $Q, R$ are specified. It means that the AF in the form (12) is robust to uncertainties in the specification of the covariances of the ME and ObE.

## 3.2 Stability of the AF

One of the main features of the AF is related to its stability.

For simplicity of presentation, in the previous section, the AF algorithm is written out under the assumption (13). In practice, application of the AF in the form (13) is not recommended since instability may occur. It is easy to see that the transition matrix of the filter is given by

$$L = (I - KH)\Phi \quad (18)$$

It is evident that if we do not take care on the structure of $K$, varying stochastically all elements of $K$ can lead to instability of $L$ and the filter will be exploded. Moreover, for HdSs, the number of elements of $K$ is very large. It is therefore primordial to choose a parametrized stabilizing structure for $K$ (depending on $\theta$) to ensure a stability of $L$ and reducing a number of tuning parameters. This question is addressed in [6]. One of possible structures for $K$ is of the form

$$K(t) = P_r \Theta K_e, K_e := M_e H_e^T \left[H_e M_e H_e^T + \alpha I\right]^{-1}, H_e := H P_r \tag{19}$$

where $P_r \in R^{nxr}$ is a matrix with dimensions $n \times r$, $r$ is the dimension of the reduced space (equal to the number of unstable eigenvectors (EiVecs)of $\Phi$), the matrix $M_e$ is a strictly positive symmetric definitive playing the role of the ECM in the reduced space $R_e$, $\Theta$ is a diagonal matrix with diagonal elements $\theta_i$ whose values belong to $(\epsilon, 2-\epsilon)$, i.e. $\theta_i \in (\epsilon, 2-\epsilon)$, with $\epsilon \in (0,1)$ whose value depends on the modulus of the first stable eigenvalue (EiV) [6]. We will refer to the filter s.t. (19) with $\Theta = Id$ ($Id$ is the identity matrix of appropriate dimension) as a *nonadaptive filter* (NAF). In the AF, the parameters $\theta_i$ are adjusted each time when a new observation arrives, to minimize the cost function (9). Thus $\theta_i$ is a time-varying function. As to the matrix $P_r$, its choice is important to ensure a filter stability. One simple and efficient procedure (called Prediction Error Sampling Procedure—PeSP) to generate $P_r$ is to use the power orthogonal iteration method [18] which allows to compute real leading Schur vectors (SchVecs) of $\Phi_i$. The advantage of using the SchVecs compared to the EiVecs, is that they are real and their computation is stable. It is seen that the optimal AF is found in a class of stable filters which is stable even for an unstable numerical model. As to the VM, the optimal trajectory is found on the basis of only the numerical model with the initial state as a control vector. It means that for unstable dynamics, the errors in the forcing or numerical errors arising during computations will be amplified and lead to large estimation error growth. More seriously, the VM requires a large set of observations and large number of iterations (i.e., many forward and backward integrations of the direct and adjoint models) which naturally leads to increase of estimation error too.

### 3.3 On improving the initial gain

Consider the gain structure (19). Suppose that $M_e$ has been chosen in agreement with the required stability conditions. Before tuning the parameters $\theta_i$ to minimize the cost function (9), remark that stability of the filter is still ensured for the following gain:

$$K(t) = P_r \Lambda \Theta K_e, \tag{20}$$

where $\Lambda = \mathrm{diag}(\lambda_1, \ldots, \lambda_r)$, $\lambda_k \in (0,1)$ since then for $\Gamma = \Lambda\Theta = \mathrm{diag}(\gamma_1, \ldots, \gamma_r)$ it implies $\gamma(t) = \lambda(t)\theta(t) \in (\epsilon, 2-\epsilon)$, where $\epsilon \in (0,1)$. Writing the equation for the filtered error (FE) $e_f(t) := \hat{x}(t) - x(t)$ one sees that the matrix $L$ in (18) also represents the transition of the FE $e_f(t)$. It means that it is possible to choose a more optimal initial gain by solving, for example, the minimization problem

$$J_o(\Lambda) = \|L(\Lambda)\|_2^2 \to min_\Lambda \tag{21}$$

The problem (21) is solved without using the observations, hence it is offline. Once the optimal $\Lambda^*$ has been found, the standard AF is implemented s.t. the filter gain

$$K(t) = P_r \Theta K_e^*, K_e^* := \Lambda^* K_e \tag{22}$$

It is seen that using the structure (22) this optimization procedure does not require the information on the ME statistics.

## 4. Joint estimation of state and model error in AF

The previous section shows how the AF is designed to deal with the difficulty in specification of covariances of the ME and ObE. This is done without exploiting a possibility to determine, more or less correctly, a subspace for the ME. If such a subspace can be determined without major difficulties, it would be beneficial for better estimating the AF gain and improving the filter performance. In [19], the hypothesis of the structure of the ME has been introduced and a number of experiments have been successfully conducted.

There is a long history of joint estimation of state and ME for filtering algorithms, in particular, with the bias and covariance estimation. One of the most original approaches, dealing with the treatment of bias in recursive filtering (known as bias-separated estimation—BSE), is carried out by Friedland in [20]. He has shown that the MMSE *state estimator* for a linear dynamical system augmented with bias states can be decomposed into three parts: (1) bias-free state estimator; (2) bias estimator; and (3) blender. This BSE approach has the advantage that it requires fewer numerical operations than the traditional augmented-state implementation and avoids numerical ill-conditioning compared to the case of bias-separated estimation by filtering technique.

It is common to treat the bias as part of the system state and then estimate the bias as well as the system state. There are two types of ME—deterministic (DME) and stochastic (SME). Generally speaking, a suitable equation can be introduced for the ME. In the presence of bias, under the assumption on constant $\boldsymbol{b}$, instead of (1) one has

$$x(t+1) = \Phi x(t) + b(t) + w(t), b(t+1) = b(t), t = 0, 1, 2 \ldots \quad (23)$$

To introduce a subspace for the variables $w(t), b(t)$ the SME and DME in (23), let

$$w = G_w \rho, b = G_b d$$
$$G_w \in R^{n \times n_w}, G_b \in R^{n \times n_d}, n \geq n_w, n \geq n_b \quad (24)$$

Generally speaking, $G_w, G_b$ are unknown, and finding reasonable hypothesizes for them is desirable but not self-evident. In [19], one hypothesis for $G_w, G_b$ has been introduced (it will be referred to as Hypothesis on model error—HME).

The information on $G_w, G_b$, given in (25), allows to better estimate the DME $b$ and SME $w$ for improving the filter performance, especially for $n_b < n$, $n_w < n$ in a HdS setting. The difficulty, encountered in practice of operational forecasting systems, is that (practically) nothing is given a priori on the space of the ME values. To overcome this difficulty, one simple hypothesis has been introduced in [19]. This hypothesis is postulated by taking into consideration the fact that for a large number of data assimilation problems in HdSs, the model time step $\delta t$ (chosen for ensuring a stability of numerical scheme and for guaranteeing a high precision of the discrete solution) is much smaller than $\Delta t$—the assimilation window (time interval between two successive observation arrivals).

Suppose that $\Delta t = n_a \delta t$ where $n_a$ is a positive integer number.

*Hypothesis* (on the subspace of ME—HME) [19]. Under the condition that $n_a$ is relatively large, the ME belongs to the subspace spanned by all unstable and neutral EiVecs (or SchVecs) of the system dynamics $\Phi$.

## 5. Simple numerical examples

### 5.1 One-dimensional system

To see the difference between the AF and the KF in doing with ME uncertainties, introduce the one-dimensional system

$$x(t+1) = \Phi x(t) + w(t), z(t+1) = hx(t+1) + v(t+1), t = 0, 1, \dots \quad (25)$$

In (25), $\Phi$ is the unique eigenvalue (also the singular value) of the system dynamics.

i. For simplicity, let $\Phi = 1, h = 1$. This corresponds to the situation when the system is *neutrally stable*. The filter fundamental matrix (18) now is $L(K) = (1 - K)$ which is stable if $K \in (0, 2)$. For the KF gain (4)–(8), as $K_{kf}(t) = \frac{M_{kf}(t)}{M_{kf}(t)+R}$ we have $K_{kf}(t) \in (0, 1)$, $M_{kf}(t)$ is the solution of (7). That is true for any $M_{kf}(t) \geq 0$, $R > 0$. It means then the KF is stable. Mention that if $Q > 0$ always $M_{kf}(t) > 0$. In general, $K_{kf}(t) = M_{kf}(t)\left[M_{kf}(t) + R\right]^{+}$ where $[A]^{+}$ is the pseudo-inverse of $A$ [21].

   For the AF, we have in this case $P_r = 1$. Consider the gain $K_{af}(\theta) := P_r \theta K_e$, where $K_e$ is the gain of the form $K_e = \frac{M_e}{M_e+R}$, $M_e > 0$, $R > 0$, $M_e$ is constant. We have then for the NAF ($\theta = 1$) $0 < K_e < 1$ and $K_{naf} = K_e$.

   For the AF, the transition matrix (18) reads $L_{af}(\theta) = (1 - \theta K_e)$. For $\theta \in (0, 2)$, $|L_{kf}(\theta)| \in (0, 1)$, $K_{af}(\theta) \in (0, 2)$ and the AF is stable. It is evident that there is a larger margin for varying the gain in the AF than that in the KF since $K_{kf}(t) \in (0, 1)$. One sees that the stationary KF is a member of the class of stable AFs (19). The performance of AF is optimized by solving the problem (9) using the procedure (11) and (12) or SPSA algorithms (*Comment 2.2*).

ii. Let $\Phi < 1$, i.e., the system (1) is *stable*. The results in (i) are valid for the AF structure. In this situation, the filter is stable even for $K_{af} = 0$.

iii. Let $\lceil \Phi \rceil > 1$—the system (1) is *unstable*. Consider two situations (a) $\Phi > 1$; (b) $\Phi < -1$. As before $P_r = 1$.

For $\Phi > 1$ we have

$$\frac{\Phi - 1}{K_e \Phi} < \theta < \frac{\Phi + 1}{K_e \Phi} \quad (26)$$

In particular, when $\Phi \to 1$, approximately $\theta \in \left(0, \frac{2}{K_e}\right)$. When $Q \gg R$ (that is usually in practice), approximately $\theta \in (0, 2)$ as in the situation (i). For large $\Phi \gg 1$, $\frac{\Phi-1}{K_e\Phi} \to \frac{1}{K_e}$ (left-hand limit), $\frac{\Phi+1}{K_e\Phi} \to \frac{1}{K_e}$ (right-hand limit) and there remains no margin for varying $\theta$ (or $Q \gg R$) and $K_{af} \to 1$. It is important to emphasize that as $K_e$ is chosen by designer, we can define the interval for varying $\theta$ if the amplitude of $\Phi$ is more or less known. In practice, one can vary $\theta \in [\epsilon, 2 + \epsilon]$ with small $\epsilon > 0$ for $\Phi$ close to 1, and with $\epsilon$ close to 1 for large $\Phi$.

For $\Phi < -1$ we have

$$\frac{\Phi + 1}{K_o \Phi} < \theta < \frac{\Phi - 1}{K_o \Phi} \tag{27}$$

It is seen from (27) that when $\Phi \to -1$, approximately $\theta \in \left(0, \frac{2}{K_e}\right)$. As for the situation $\Phi \ll -1$, $\frac{\Phi+1}{K_o \Phi} \to \frac{1}{K_o}$ (left-hand limit), $\frac{\Phi-1}{K_o \Phi} \to \frac{1}{K_o}$ (right-hand limit) when $Q \gg R$ approximately $\theta \to \frac{1}{K_o}$ hence $K_{af} \to 1$.

It is important to stress that the KF gain is computed on the basis of $Q$ and $R$ (under the condition that the statistics of the initial state will be forgotten as $t$ becomes large); whereas, the gain of the AF is updated on the basis of samples of the innovation vector. It means that the KF is optimal in the MMSE sense (under the condition of exact knowledge of the required statistics) whereas the AF is optimized during the assimilation process using PE realizations of the system output (innovation vector). The KF gain can be computed in an offline fashion, whereas the AF gain is a function of observation and computed in online.

## 5.2 Two-dimensional system: specification of covariances

### 5.2.1 Stable filter

To see the role of the correction subspace $R[P_r]$ in ensuring a stability of the AF, let us consider the system (1) and (2) s.t.

$$\Phi = diag\ (\Phi_{11}, \Phi_{22}), H = diag\ (1, 1) \tag{28}$$

Consider the AF with the gain (19) s.t. $P_r = Id$, i.e., two columns of $P_r$ are in fact the EiVecs associated with two EiVs $\Phi_{11}$ and $\Phi_{22}$. Let us denote the AF gain $K_{af} = P_r \Theta M H^T \left[HMH^T + R\right]^{-1} = \Theta K_e$ with $K_e := M_e H^T \left[H M_e H^T + R\right]^{-1}$ which is structurally identical to that of the KF. For the nonadaptive filter $K_{naf} = K_e$ and with the choice $M_e = diag\ (M_{11}, M_{22})$, taking into account (28) one gets

$$K_{naf} = diag\ (K_{11}, K_{22}), K_{ii} = \frac{M_{ii}}{M_{ii} + R_i}, i = 1, 2 \tag{29}$$

$$L_{naf} = diag(l_{11}, l_{22}), l_{ii} = \left(1 - \frac{M_{ii}}{M_{ii} + R_i}\right)\Phi_{/ii}, i = 1, 2 \tag{30}$$

The filter transition matrix (30) is obtained on the basis of $L_{naf} = (I - K_{naf} H)\Phi$ and the assumption (29). It is easily to see that $L_{naf}$ has two EiVs, $\lambda_i = l_{ii}, i = 1, 2$ where $l_{ij}$ est. the $(ij)$ element of $L_{naf}$.

Stability of the filter depends on the condition $|l_{ii}| < 1, i = 1, 2$. For $M_{ii} > 0, R_i > 0$, if $\Phi_{ii}$ is *stable* or *neutrally stable*, i.e. $|\Phi_{ii}| \le 1, i = 1, 2$, we have $|l_{ii}| < 1$. For *unstable* $\Phi_{ii}$ ($|\Phi_{ii}| > 1, i = 1, 2$), the filter is unstable if $\Phi_{ii} > \frac{M_{ii}+R_i}{R_i}$ (situation $\Phi_{ii} > 1$) or $\Phi_{ii} < -\frac{M_{ii}+R_i}{R_i}$ (situation $\Phi_{ii} < -1$). These conditions should be taken into account when the EiVs of $\Phi$ are large.

For the AF gain (19) ($P_r = I$),

$$l_{ii} = \left(1 - \frac{\theta_i M_{ii}}{M_{ii} + R_i}\right)\Phi_{ii}, \tag{31}$$

From (31) conditions for $|l_{ii}| < 1$ can be obtained as done in Section 5.1 with the one-dimensional system since $l_{ii}, i = 1, 2$ are independent one from another. The length of the interval $I_i$for varying $\theta_i$ depends on the value of $\Phi_{ii}$ (see (26)).

This example shows that for $P_r = Id$, it is always possible to construct a stable AF whatever are the EiVs of $\Phi$ (stable or unstable). There are some constraints for $M_{ii}$ (they are positive) and for $R_i$ (small positive). Optimality of the AF is obtained by searching recursively (in time) the optimal $\theta_i$ during assimilation process. Thus, in the AF, a correct specification of ME and ObE statistics (second order) is not important as happens in the KF.

### *5.2.2 Unstable filter*

Consider the situation when $P_r$ is constructed from only one vector. Let $P_r = (1, 0)^T$—the EiVec associated with $\Phi_{11}$ (the results remain the same if we choose $P_r = (0, 1)^T$—the EiVec associated with $\Phi_{22}$).

$$\Phi = diag\ (\Phi_{11}, \Phi_{22}), |\Phi_{11}| > 1,\ |\Phi_{22}| > 1, H = diag(1, 1) \tag{32}$$

We show now that the filter with the gain (19) is unstable. We have (for $\Theta = Id$),

$$H_e = HP_r = P_r, K = P_r K_e, K_e = M_e H_e^T \left[H_e M_e H_e^T + R\right]^{-1} = M_e P_r \left[P_r M_e P_r^T + \alpha I\right]^{-1} \tag{33}$$

if we put $R = \alpha I$. For $L_{naf} = [I - KH]\Phi$, taking into account (32), (33) it implies

$$L_{naf} = \begin{bmatrix} \dfrac{\alpha \Phi_{11}}{m_e + \alpha} & 0 \\ 0 & \Phi_{22} \end{bmatrix} \tag{34}$$

As $\frac{\alpha}{m_e + \alpha}$ can be made as small as desired by choosing small $\alpha > 0$, the first EiV $l_{11} = \frac{\alpha \Phi_{11}}{m_e + \alpha}$ can be made stable. However, the second EiV in (34) $l_{22} = \Phi_{22} > 1$ is unstable. It implies that the filter with the gain (19) s.t. $P_r = (1, 0)^T$ is unstable. This happens even for $\Theta \neq Id$. It means that when the projection subspace $R[P_r]$ does not contain all unstable and neutral EiVecs of the system dynamics, it is impossible to guarantee a stability of the filter.

## 5.3 Two-dimensional system: estimation of ME

Consider the filtering problem (1) and (2), the dynamical system (1) describes a sequence of system states at time instants $t = 0, 1, \ldots$ when the observations are available. It means that $\Phi$ represents the transition of system state over the (observation) time window $\Delta t$ separating arrivals of two successive observations. In practice, the interval $\Delta t$ is much larger than the *model time step* $\delta t$ which is the step size in approximating the temporal derivative. The choice of $\delta t$ is important for guaranteeing a stability of discretized scheme and having high is important for guaranteeing a stability of discretized scheme and having high precision of the discretized solution (wrt the continuous solution). We have then $\Delta T = n_a \delta t$, where $n_a$ is a relatively large positive integer. For example, in the HYCOM model at SHOM (French marine) for the Bay of Biscay configuration, the interval $\Delta t$ between two observation arrivals is 7 days which is equivalent to integrating 1200 model time steps $\delta t$. It means $n_a = 1200$. Symbolically we have then the equations for model time step integration

$$x'(\tau+1) = \Phi' x'(\tau) + \psi'(\tau), \psi'(\tau) := b'(\tau) + w'(\tau) \tag{35}$$

In (35), $\Phi'$ represents the integration of numerical model over one model time step $\delta t$. Hence

$$\Phi = [\Phi']^{n_a} \tag{36}$$

The contribution of $\psi'(\tau)$, over the assimilation window $[t-1, t]$ (for simplicity and without loss of generality, one supposes $t - 1 := 0, t := n\_a$) is

$$\psi(t) = b(t) + w(t), b(t) := \sum_{\tau=0}^{n_a} \nu_1(\tau)_1, \nu_1(\tau) := [\Phi']^{n_a - 1 - \tau} b'(\tau),$$

$$w(t) := \sum_{\tau=0}^{n_a} \nu_2(\tau),\ \nu_2(\tau) := \Phi^{n_a - 1 - \tau} w'(\tau), [\Phi']^{-1} := 0 \tag{37}$$

The HME in Section 4 says that the SME $w(t)$ and DME $b(t)$, as functions of $n_a$, belong to the subspace spanned by leading EiVs (or SchVecs) of $\Phi$ for a relatively large $n_a$. The initial filtering problem now has the form (1) and (2) s.t. (36) and (37) where $t = \tau / n_a$.

To illustrate this HME, continue the two-dimensional system in Section 5.2.2 and suppose that $|\Phi'_{11}| > 1, |\Phi'_{22}| < 1$. Applying HME in this case is equivalent to saying that the values of MEs $b(t), w(t)$, approximately, belong to the subspace $R[u_1]$ spanned by the first EiVec $u_1 = col(1, 0)$, associated with the EiV $\Phi'_{11}$. Here $y = col(y_1, \dots, y_n)$ denotes the vector-column with components $y_1, \dots, y_n$. It follows that the covariance matrix of $w(t)$ is assumed to be of the form $Q = \sigma_w^2 u_1 u_1^T$ and $b(t)$—of the structure $b(t) = c u_1$, $c$ is a scalar to be estimated. For the algorithm of joint estimation of state and bias (in term of $c$), see [19].

## 6. Simulation results

### 6.1 One-dimensional system

In this section, the filtering problem (25) in Section 5.1 is considered s.t.

$$\Phi = 0.99, H = 1.02, Q = 0.09, R = 0.01.$$

The true system states and observations are simulated using the initial state $x(0) = 1$ and $w(t), v(t)$ are zero mean Gaussian mutually uncorrelated and temporal uncorrelated sequences.

To see the performance of the AF, unknown system states are estimated on the basis of the AF algorithm. To obtain a reference, the standard KF is also implemented for solving this filtering problem. In the filtering algorithms, the estimate of the initial state is $\hat{x}(0) = 2$. The gain $K_{naf}$ in the NAF is taken as that of the KF at $t = 0$, i.e., $K_{naf} = K_{kf}(0)$.

**Figure 1** shows the temporal evolution of the parameters $\theta_m(t)$ during assimilation process.

The gains in the KF and AF during the assimilation process are displayed in **Figure 2**. Mention that the KF gain is computed s.t. true statistics $Q, R$. In the AF, $\theta_m(t)$ has been used for computation of the AF gain, i.e., $K_{af} = \theta_m(t) K$. From **Figure 2**, one sees that initialized by the same value, the two gains become different during assimilation process. The KF gain has reached a stationary regime very quickly.

The mean temporal RMS (root mean square) of the innovation is shown in **Figure 3**. It is interesting to remark that no significant difference is observed between two curves and a slightly better performance is produced by the KF.

In **Figure 4**, we show RMS of the state FE produced by the KF and AF under the condition that the variance $Q$ is known exactly. One sees that the KF, as expected, produces the best results.

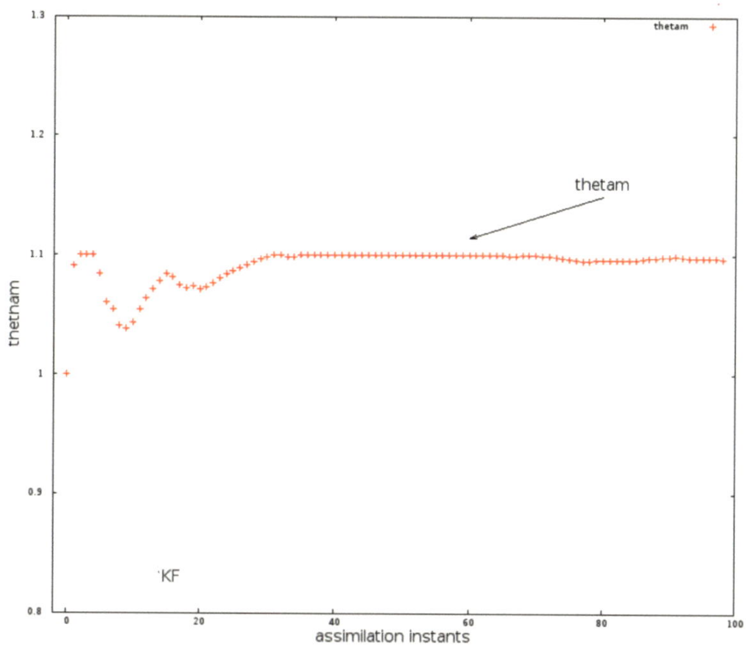


**Figure 1.**
*Temporal evolution of the parameter $\theta_m(t)$ in the AF gain.*

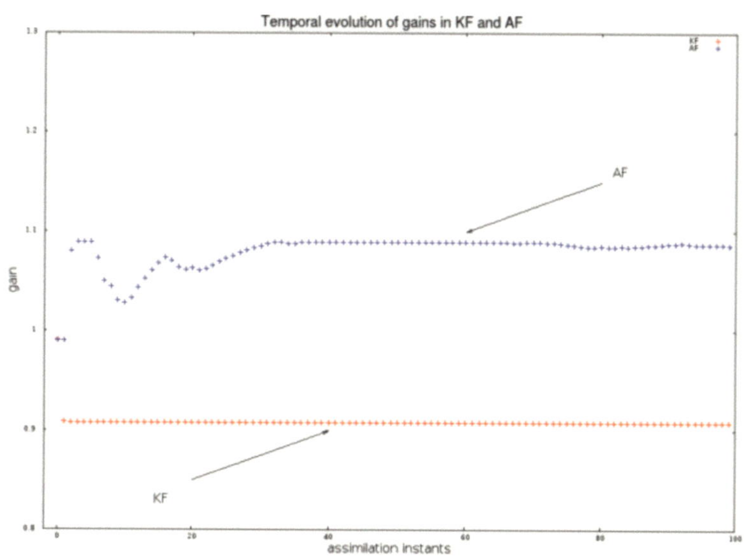


**Figure 2.**
*Temporal evolution of gains in KF and AF during data assimilation.*

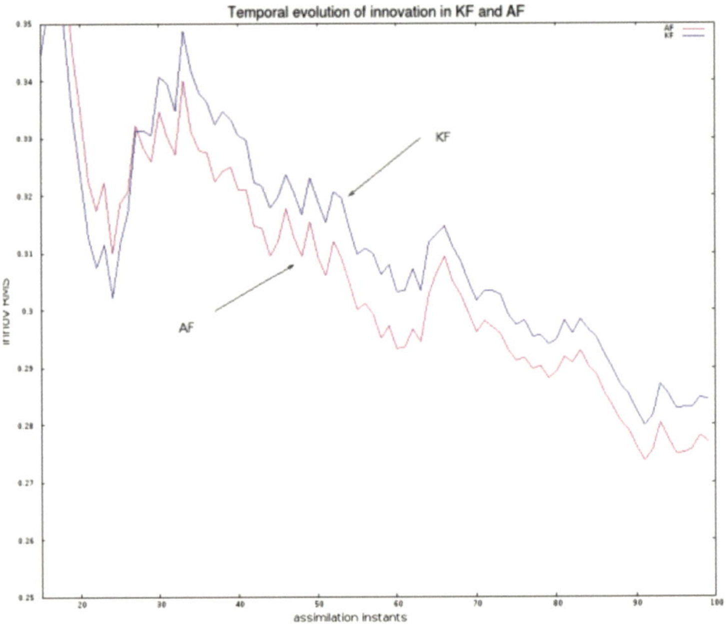


**Figure 3.**
*RMS of innovation produced by the KF and AF.*

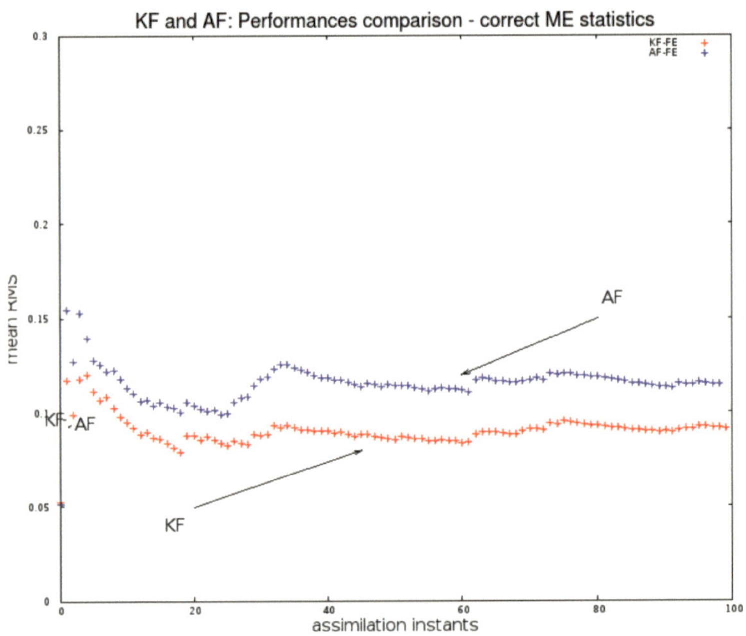


**Figure 4.**
*RMS of the state FE produced by the KF and AF under the condition that the variance of ME is known exactly. It is seen that when the ME is correctly specified, the KF behaves better than the AF.*

**Figure 5** shows the RMS of FE as a function of the variance $Q$. Here, the value of $Q$ varies from 0.1 to 1.9. Note that the true value of $Q$ is 0.1. The red curve represents the RMS of FE produced by the KF at the end of the assimilation period (as a function of $Q$). The green curve has the same meaning, but for the FE

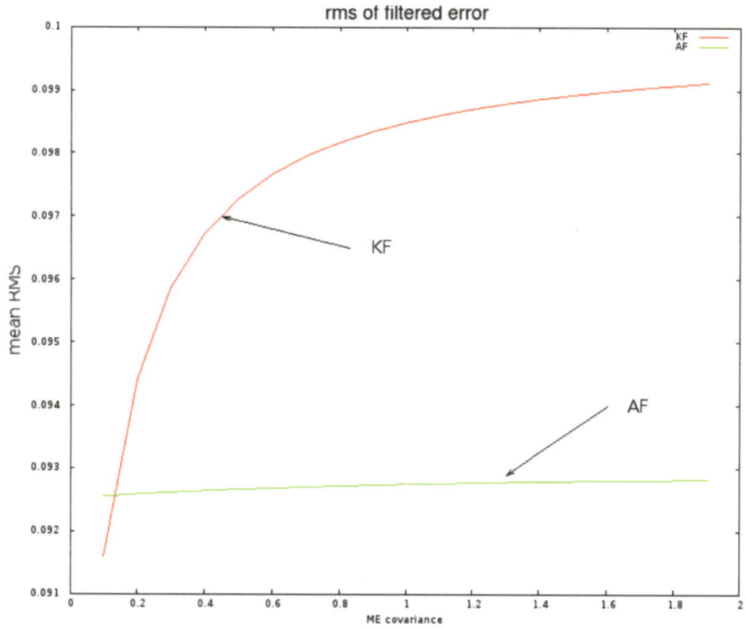


**Figure 5.**
*RMS of FE as a function of Q. The true value of Q is equal to 0. It is noted that the KF behaves better than the AF s.t. true Q but is more and more degraded as the ME becomes greater and greater. At the same time, the FE of the AF remains very robust.*

produced by the AF. It is interesting to note that when $Q$ is correctly specified, the KF behaves better than the AF, but misspecification of $Q$ leads to growing of the error in the KF. The AF is robust wrt the error in the specification of $Q$. This fact says in favor of the AF as an efficient tool for overcoming uncertainties in the ME.

## 6.2 Two dimensional system

### *6.2.1 Illustration of hypothesis HME*

According to the notations in Section 5.3, consider the two-dimensional system (1) with $\Phi' = [\Phi'_{ij}], \Phi'_{11} = 1.02, \Phi'_{12} = 0.1, \Phi'_{21} = 0, \Phi'_{22} = 0.9, H = (1, 1)$ with the true DME $b'(\tau) = \text{col}(0.1, 0.1)'$. Thus the first EiV is unstable, the second—stable [19].

Numerically one finds that the first SchVec is equal to $u_1 = (-1, -7.0\,E - 7)^T$.

**Figure 6** [19] shows the simulation results obtained on the basis of (37). One sees that, for $n_a > 10$, the second component of $w(t)$ is close to 0 whereas the first component becomes bigger and bigger (in absolute value) as $n_a$ increases. Here, $w'(\tau)$ is a sequence of independent two-dimensional Gaussian random vectors of zero mean and variance 1. This means that the values of $w(t)$ become more and more close to the subspace $R[u_1]$ spanned by $u_1$, hence the HME is practically valid for $n_a > 10$ in this example. Mention that, as a rule, in ocean numerical models, $n_a$ is of order $o(100)$ ($n_a = 800$ or the MICOM model in the experiment in Section 6.3). See also [22].

### *6.2.2 Assimilation results*

First simulation of the sequence of true system states $x'(\tau), \tau = 1, \dots, 390$ has been carried out (see (35)). The observations are picked at $\tau = 15, 30, \dots, 390$.

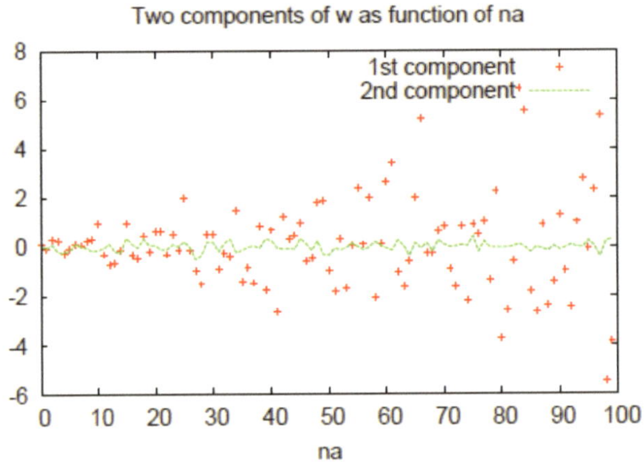


**Figure 6.**
*Two components of $w(t)$ as functions of $n_a$.*

In terms of $x(t)$, the filtering problem then is of the form (1) and (2) s.t. $t = \frac{\tau}{15}, \Phi = \Phi'^{15}$ (if no bias exists). When there is a bias, instead of $w(t)$ stands $\psi(t) := b(t) + w(t)$.

In the experiment, the true system states $x'(t)$ are generated by (35) s.t. $b'(\tau) = col(0.1, 0.1), w'(t)$ is zero mean with the covariance $Q' = Id$. The observation error is of zero mean and covariance $R = 0.16$. For the state $x(t)$ in the KF and NAF, the forecast is obtained at each assimilation instant $t$ as $\hat{x}(t+1/t) = \Phi\hat{x}(t) + \hat{b}(t)$. The simulation yields $b(t) = (0.2296, 2.0589\text{E} - 02)$ which results from applying (37) s.t. $b'(\tau) = col(0.1, 0.1)$.

**Figure 7** depicts the time evolution of the KF and AF gains. One sees here as in the experiment with 1D system (**Figure 2**) that the KF gain is stabilized very quickly compared to that of the AF gain.

**Figure 8** (from [19]) shows the sample time average RMS of the state FE produced by the three filters NAF, KF, and AF. One sees that the AF outperforms the NAF and KF.

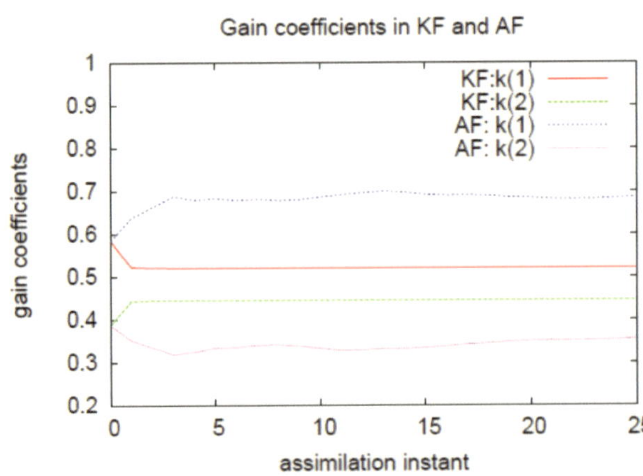


**Figure 7.**
*Gain coefficients in KF and AF: The gains in KF and AF are identical at the beginning of the assimilation process.*

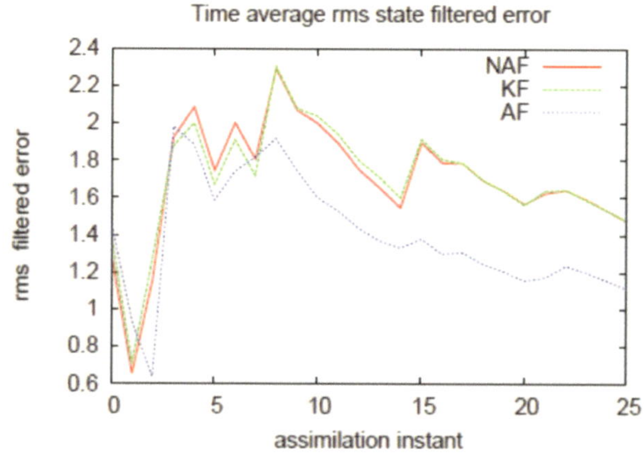


**Figure 8.**
*Sample time average RMS of the state filtered error produced by the NAF, KF, and AF.*

### 6.3 Data assimilation in the high-dimensional ocean model

To illustrate the effectiveness of the AF in dealing with uncertainties in HdSs, this section presents the results on data assimilation in the oceanic numerical model MICOM (Miami Isopycnic Coordinate Ocean Model) [19]. This MICOM describes the oceanic circulation in the North Atlantics. The model has four vertical layers with the state consisting of three variables $x = (h, u, v)$ where $h$ is layer thickness, $(u, v)$ are two velocity components. The horizontal grid is $140 \times 180$. Totally at each time instant $t$ we have the state $x(t)$ of dimension 302400 $(140 \times 180 \times 4\ (layers) \times 3\ (variables))$ . For more details on the configuration of this experiment, see [22].

The experiment is carried out on estimating the oceanic circulation using sea surface height (SSH) measurements. The SSH observation is available each 7 days $(ds)$ (hence the observation window $\Delta T = 7ds$). Mention that simulating the circulation over 7 $ds$ requires 800 model time steps $(\delta t)$ integration.

#### *6.3.1 AF with optimal initial gain*

First, in order to examine whether the method of optimal gain initialization, described in Section 3.2, is really useful for improving the filter performance, the optimization problem (21) has been solved. Symbolically, in the gain (20), $P_r = \left[Id, (QG)^T\right]^T$ where $Id$ is the identity operator on the space of layer thickness $h$, $QG$ is the quasi-geostrophy operator computing the correction for velocity using the SSH innovation $\zeta$. The gain $K_e$ computes the correction for $h$ using $\zeta$. The ECM $M = M_v \otimes M_h$—Kronecker product of $M_h$—ECM of horizontal variable, $M_v$—ECM with vertical variable (see below for details). The two parameter matrices $\Theta$ and $\Lambda$ are related to parameterization of $M_v$. The problem (21) is solved s.t. $\Theta = Id$—identity operator.

The optimal parameters $\lambda_i, i = 1, \dots, 4$ are found by solving the minimization problem (21) using SPSA algorithm. **Figure 9** shows the averaged values (see *Comment 2.1*) of $\lambda_i, i = 1, \dots, 4$ resulting during the optimization process. All $\lambda_i, i = 1, \dots, 4$ are initialized as $\lambda_i = 1, i = 1, \dots, 4$.

The two NAFs are performed, one (denoted as NAFI) is with the gain (20) s.t. $\Theta = Id$, $\Lambda = Id$, and the other (denoted by NAFOI) s.t. $\Theta = Id$ and $\lambda_i, i = 1, \dots, 4$ obtained by solving (21) (their values are those displayed at the end of the

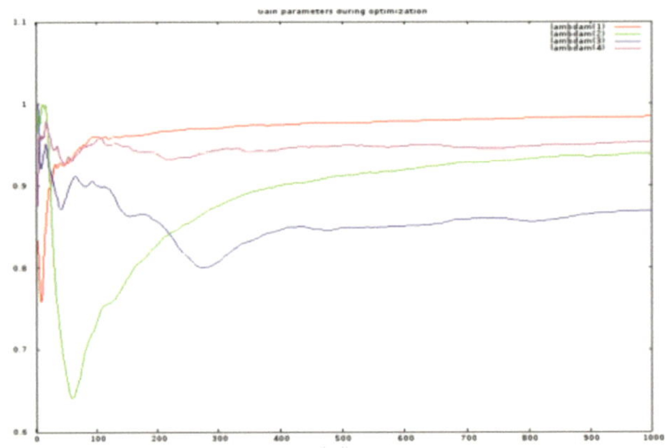


**Figure 9.**
*Control parameters $\lambda_i, i = 1, \ldots, 4$ during optimization process.*

optimization process in **Figure 9**). The performances of these two NAFs are shown in **Figure 10**. One sees here that the NAFOI has improved considerably the quality of estimates of the velocity $u$-component compared with the NAFI. This result justifies that offline optimization (21) is an interesting strategy for finding the optimal initial gain in the NAF.

### *6.3.2 Estimating the ECM of ME*

In practice, for real operational systems, information on the space of ME is not available or very poorly known. Usually, there is a big difference between the model and the real physical process and if the ME statistics are taken more or less properly, in some way, in the filtering algorithm, one can improve the filter performance and reduce the estimation error.

This idea is tested here by applying the HME in Section 4. We carry out the procedure for estimating the ECM of the ME by first constructing the subspace for the ME. For more details on the structure of the ECM $M$ in the AF, see [23]. According to [23], the ECM $M$ is assumed to be of the structure $M = M_v \otimes M_h$—the Kronecker product of $M_h$ with $M_v$ where $M_h$ is the ECM of the horizontal variable, $M_v$—ECM with vertical variable. **Figure 11** displays RMS of FE for the $u$ velocity

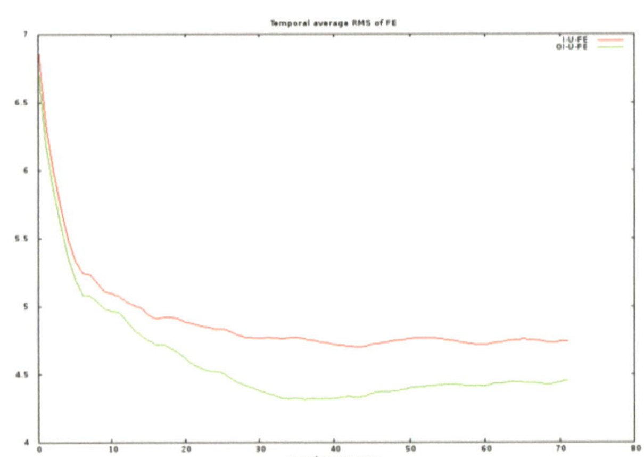


**Figure 10.**
*Temporal average RMS of FE for total velocity u-component produced by the NAF s.t. the initial gain (red curve) and the NAF s.t. optimal initial gain resulting from solving (22) (green curve).*

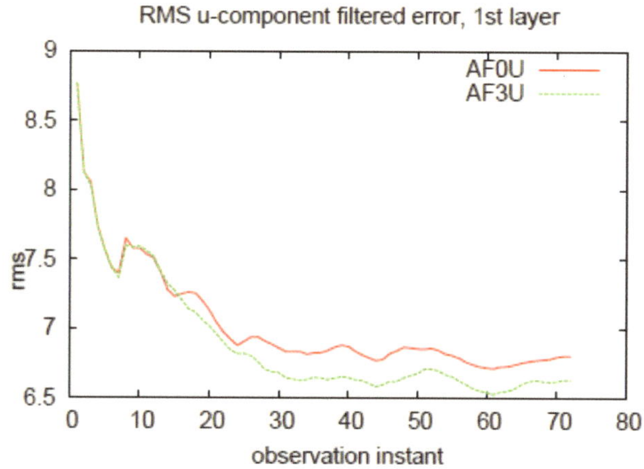


**Figure 11.**
*Performance of the AF: (i) AF0U—no ME ECM has been taken into account; (ii) AF3U—with ME ECM computed in accordance with the HME.*

component at the surface resulting from two AFs. The curve $AF0U$ corresponds to the AF whose nonadaptive version has the gain computed on the basis of the ECM $M$ using an ensemble of PE samples (generated by the PeSP in [18]). The curve $AF3U$ shows the performance of the AF with the modified ECM (by adding the vertical ME covariance $Q_v$ to the vertical ECM $M_v$). More precisely, $Q_v$ is assumed to belong to the subspace spanned by three leading EiVs of $M_v$. This choice is justified by the fact: the eigenvalue decomposition of $M_v$ has the first three EiVs with the explained variances 67, 17, 15%, respectively. As the fourth EiVec has only the explained variance 0.7E-07%, it is dropped from the subspace constructed for the vertical ME. The better performance of the $AF3U$, in comparison with that of the $AF0U$, is apparently seen in **Figure 11**.

The above experiment shows in details how, on the basis of HME, the subspace for the ME can be constructed, and how one estimates the ECM for the model error. The superior performance of the $AF3U$ over that of the $AF0U$ validates the usefulness of the HME which can serve as an important tool for estimating the ME and improving the performance of the AF for solving the data assimilation problems with HdSs.

## 7. Conclusions

One of the key assumptions to ensure the optimal performance of the KF is that a priori knowledge of the system model is given without any uncertainty. This assumption, however, is never valid in practice for dynamical systems under consideration. The uncertainties exist everywhere in modeling a real process like structural uncertainty, model parameterization, model resolution, model bias or ME statistics. For HdSs, order reduction introduced either in the original numerical model or in the filtering algorithms, inevitably leads to uncertainty in the ME, especially in geophysical numerical models.

Our focus in this chapter is to show how the AF solves efficiently filtering problems for systems operating in an uncertain environment.

As seen from this chapter, the AF has proven to be efficient to deal with uncertainties in the specification of the ME statistics, system bias or model reduction. The reasons of the success of the AF are that (i) it belongs to the class of parametrized stable filters; (ii) it is defined as the best member minimizing mean PE for the system outputs; (iii) The tuning parameters are chosen as elements of stabilizing gain and they are of no physical sense.

It is obvious from this chapter that the performance of the AF is comparable with that of the KF when perfect knowledge of all ME statistics is given, and it outperforms the KF in presence of uncertainties. This happens since the AF acquires knowledge during assimilation process, regardless of uncertainties existing in the filtering problems. From the computational point of view, implementation of the AF consumes much less memory and computational time than the KF or other assimilation methods.

Simple numerical examples and simulation results, presented in Sections 5 and 6, clearly demonstrate the advantages gained through application of the AF in dealing with uncertainties. These positive results encourage a wide application of the AF in different fields of technology and applied sciences like automatic control, finance, aerospace, space exploration, meteorology, and oceanography. A more in-depth and significant research on the capacity of the AF to deal with uncertainties is surely a challenge for the near future.

## Author details

Hong Son Hoang* and Remy Baraille
REC/HOM/SHOM, Toulouse, France

*Address all correspondence to: hhoang@shom.fr

## References

[1] Kalman REA. New approach to linear filtering and prediction problems. Journal of Basic Engineering. 1960;**82**: 35-45. DOI: 10.1115/1.3662552

[2] Kailath T, Sayed AH, Hassibi B. Linear Estimation. NJ, Upper Saddle River: Prentice-Hall; 2000

[3] Liptser RS, Shiryaev AN. Statistics of Random Processes—I. General Theory. Berlin and Heidelberg: Springer-Verlag; 2001

[4] Sayed AH. Fundamentals of Adaptive Filtering. NJ: Wiley; 2003

[5] Kucera V. The discrete Riccati equation of optimal control. Kybernetika. 1972;**8**(5):430-447

[6] Hoang HS, Talagrand O, Baraille R. On the design of a stable adaptive filter for high dimensional systems. Automatica. 2001;**37**:341-359

[7] Simon D. Optimal State Estimation. Hoboken, NJ: John Wiley and Sons; 2006. ISBN: 978-0-471-70858-2

[8] Gustafsson F, Hendeby G. Some relations between extended and unscented Kalman filters. IEEE Transactions on Signal Processing. 2012; **60**(2):545-555

[9] Evensen G. The ensemble Kalman filter: Theoretical formulation and practical implementation. Ocean Dynamics. 2003;**53**(4):343-367. DOI: 10.1007/s10236-003-0036-9

[10] Chen Y, Snyder C. Assimilating vortex position with an ensemble Kalman filter. Monthly Weather Review. 2007;**135**(5):1828-1845. DOI: 10.1175/MWR3351.1

[11] Del Moral P. Non linear filtering: Interacting particle solution. Markov Processes and Related Fields. 1996;**2**(4): 555-580

[12] Fitzgerald R. Divergence of the Kalman filter. IEEE Transactions on Automatic Control. 1971;**16**(6):736-747

[13] Hoang HS, De Mey P, Talagrand O, Baraille R. A new reduced-order adaptive filter for state estimation in high dimensional systems. Automatica. 1997;**33**(8):1475-1498

[14] Polyak BT. New method of stochastic approximation type. Automation and Remote Control. 1990; **51**(7):937-946

[15] Spall JC. Introduction to Stochastic Search and Optimization. New Jersey: Wiley; 2003. ISBN 978-0-471-33052-3

[16] Hoang HS, Baraille R. Stochastic simultaneous perturbation as powerful method for state and parameter estimation in high dimensional systems. In: Baswell AR, editor. Advances in Mathematics Research. Nova Science Publishers; 2015;**20**:117-148. ISBN: 978-1-63482-741-6c

[17] Hoang HS, Baraille R. A comparison study on performance of an adaptive filter with other estimation methods for state estimation in high-dimensional system. In: Hokimoto T, editor. Advances in Statistical Methodologies and their Application to Real Problems. Rijeka, Croatia: IntechOpen; 2017. pp. 29-52. DOI: 10.5772/67005

[18] Hoang HS, Baraille R. Prediction error sampling procedure based on dominant Schur decomposition. Application to state estimation in high dimensional oceanic model. Journal of Applied Mathematics and Computing. 2011;**218**(7):3689-3709

[19] Hoang HS, Baraille R. On estimation of model error by an adaptive filter. WSEAS Transactions on Systems and Control. 2019;**14**:158-168

[20] B. Friedland B. Treatment of bias in recursive filtering. IEEE Transactions on Automatic Control. 1969;**14**:359-367

[21] Albert AE. Regression and the Moore-Penrose Pseudo-Inverse. London: AP; 1972

[22] Hoang HS, Baraille R, Talagrand O. On an adaptive filter for altimetric data assimilation and its application to a primitive equation model, MICOM. Tellus 57A; 2005:153-170. DOI: 10.1111/j.1600-0870.2005.00094.x

[23] Hoang HS, Baraille R. On the efficient low cost procedure for estimation of high-dimensional prediction error covariance matrices. Automatica. 2017;**83**:317-330. DOI: 10.1016/j.automatica.2017.06.018

Chapter 3

# Convolutional Neural Network Demystified for a Comprehensive Learning with Industrial Application

*Anand Raju and Shanthi Thirunavukkarasu*


## Abstract

In the recent past of time, numerous investigators have driven on and subsidized novelties to image classification methods. In this chapter, an introduction to image classification scheme and their types is offered. Image classification discovers its application in a variety of fields, to name a few, judgment of diseases, finding and identification of faults, classification of nutrition goods based on superiority, valuation of usual capitals and conservation pollution, education of land use and land cover from remote sensing satellite images, character identification and detection in optical character reader, face recognition, object detection, and so on. Automatic image classification schemes found on actual algorithms deliver high accuracy and exactness in recognizing object/features. Convolution neural network is a superior genre of neural network that requires minimal preprocessing. The ability of the convolutional neural network (CNN) to understand the visual content of the input image makes its suitable for recognizing minute variation between the classes. This power of the CNN makes it a good choice to address image classification problems with multi-classes. So, in this chapter, the entire flow of CNN's architecture with different industrial applications will be discussed.


**Keywords:** convolutional neural network, machine learning, deep learning, python, data prepossessing, pooling, layers, architectures

## 1. Introduction

The convolutional neural network also termed as ConvNet or CNN is a form of deep learning neural network. Most of the inner layers apply mathematical convolution operation to compute the feature maps for the next layer and hence the name convolutional neural network. The effortless construction of such network makes itself useful for a variety of real-time applications like image classification systems, image recommender systems, optical character recognition (OCR) systems, medical diagnosing systems, fault detection systems, and so on. The CNN architecture is highly inspired by a pattern of neuron connectivity in human visual cortex. The minimum intervention of human in feature selection and nominal preprocessing makes itself preferable for different applications. Like the traditional neural network, CNN consists of input layer, hidden layers, and an output layer [1].

The hidden layers are formed by combinations of convolution layers, ReLU layers, and pooling layers. The CNN differs from regular neural network in several ways. The input is a three-dimensional data that usually includes the color channels of the input image. The parameter sharing and confined connectivity are unique features of CNN. Parameter sharing refers to the sharing of equal weights for all neurons in computation of a feature map. Confined connectivity enables the neurons to get connected to a specific subset of the input layer. This reduces the overfitting problem. These two features of CNN reduce the total number of learnable parameters, thereby decreasing the computation time [2, 3]. The following section explains the functions of each layer in detail.

## 1.1 Convolutional layer

The convolutional layer computes a simple dot product between the selected region of the image and set of kernel functions called filters. Generally, an image of dimension M × N × C is given as input. M × N represents the length and width of the image, and C represents the number of color channels usually C = 3 (red, green, and blue color channels) [4]. The output of the convolutional layer is called feature map and its dimensions are given as W × Q × K. The values of W and Q can be determined using the equation given below. Several parameters like filter size (F), zero padding (ZP), stride (S), and number of filters (K) are used to compute the dimensions of the feature map.

$$W = \frac{(M - F + (2ZP))}{S} + 1 \tag{1}$$

$$W = \frac{(N - F + (2ZP))}{S} + 1 \tag{2}$$

The filter parameters are explained as below filter size (F) denotes the size of the filter or kernel used for convolution operation. Generally, kernel functions are odd-numbered square matrix, i.e., 3 × 3, 5 × 5, etc. [5]; an odd value is preferred so that the center of the kernel matrix can be fixed on the pixel on which it is operated. The figure shows the convolution of 5 × 5 input image (blue-colored matrix) with 3 × 3 kernel function (green-colored matrix) and a 3 × 3 feature map output (**Figure 1**).

## 1.2 Zero padding

Zero padding refers to the number of zeroes that are added along the brim of the input matrix. The dimensions of output matrix of the convolutional layer will be

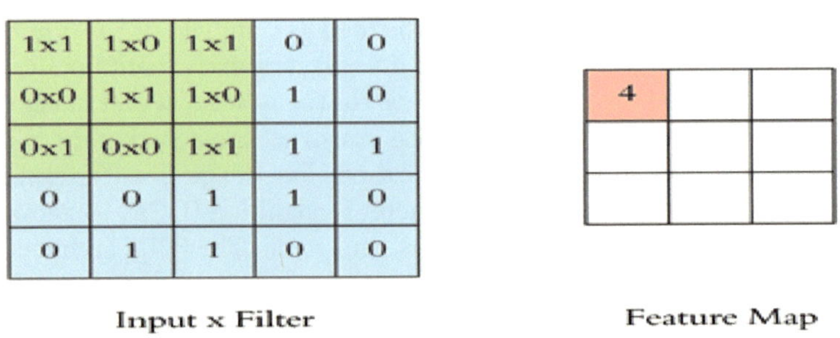


**Figure 1.**
*Convolution of 3 × 3 kernel function over a 5 × 5 input.*

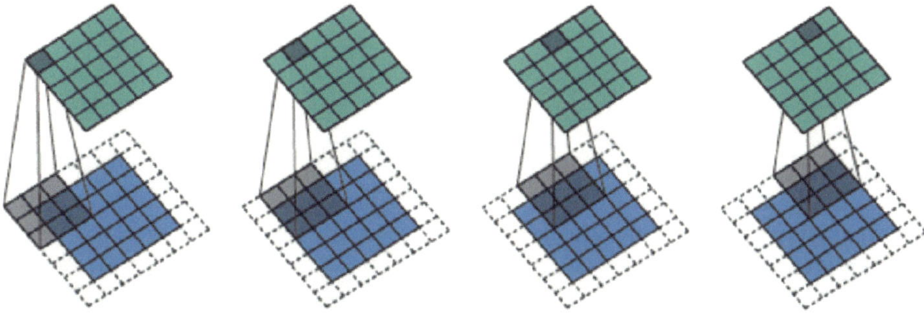

**Figure 2.**
*Demonstration of zero padding to maintain the output size same as that of the input.*

reduced because of mathematical operation performed in that layer. Appending zeroes to the input matrix will prevent the shrinkage of size of feature maps produced at the output. Maintaining the size of feature maps can be achieved by assigning zero padding as depicted in the figure, and hence, it becomes essential in networks with more number of layers (**Figure 2**).

### 1.3 Stride (S)

A filter or the kernel matrix has to be translated throughout the input matrix vertically from top to bottom and also horizontally from left to right, covering all the elements in the input matrix. Stride controls the movement of the translation. It represents the number of steps taken by the kernel matrix during its translational movement. The figure illustrates the movement of the filter when a 3 × 3 kernel function is applied over a 7 × 7 input with stride S = 2 output and results in a size of 3 × 3 (**Figure 3**).

A demonstration of the function of convolutional layer is displayed in the figure. The three-dimensional input with the volume 5 × 5 × 3 (M = N = 5, C = 3) when padded with zero outer border turns into a volume of 7 × 7 × 3 and is shown in blue color. Two filters with the volume 3 × 3 × 3 (F = 3, K = 2) are shown in red color. The three layers are shown one below the other. The filter is convolved with the input matrix with zeros in the outer border (zero padding ZP = 1) and stride S = 2. With these parameters, the dimension of the feature map is computed using the equation:

$$W = Q = \frac{(M - F + (2ZP))}{S} + 1 = \frac{(5 - 3 + (21))}{2} + 1 = 3 \quad (3)$$

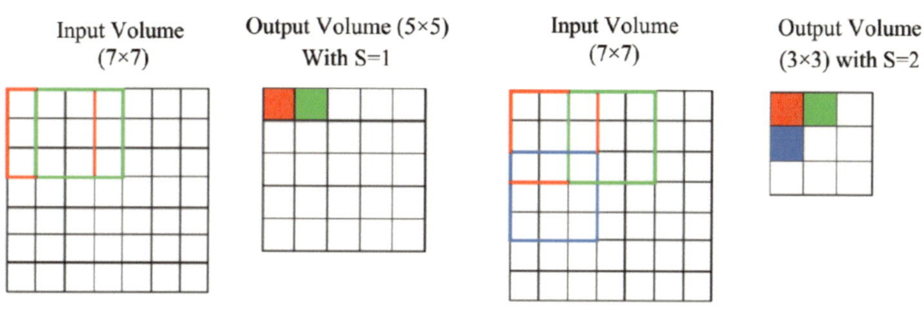


**Figure 3.**
*Application of 3 × 3 kernel function over a 7 × 7 input with stride S = 1 and S = 2.*

The dimension of the feature map is given as W × Q × K = 3 × 3 × 2, and it is shown in green color. The kernel matrix contains different weights in order to differentiate features of higher importance with other features. Filter weights are the values of the kernel matrix. Apart from assigning weights, bias values are also provided to mention the probability of occurrence of a particular pattern. The weights and biases refer to the parameters of the network. The parameters are learned by the network during the training phase, and all other parameters are termed as hyper parameters (**Figure 4**). The values in the output matrix are calculated using the equation:

$$\text{Output} = (\text{Inputxfilter}) + \text{bias} \tag{4}$$

A unique feature of CNNs is that same filter is shared with many neurons. Using the same bias and weight vectors along the region reduces the memory requirement to a great extent. The convolution explained in this section is basically a 2D convolution. It is called 2D convolution because the kernel is moved in only two directions, i.e., along the height and width of the input layer. 3D convolution is also possible if a 3D kernel is applied and moved along all three dimensions, i.e., along the height, width, and depth of the input volume. Apart from the regular convolution, there are few other types of convolutions as discussed below.

## 1.4 Types of convolution

### *1.4.1 Transposed convolution*

This type of convolution is preferred for deconvolution operation and generally introduces checkerboard effects in the output image. This type of convolution generally increases the size of output image as it does up sampling process [4].

### *1.4.2 Dilated convolution*

In this type of convolution (d-1), spaces are inserted in between the kernels before applying convolution. If d = 1, it resembles regular convolution.

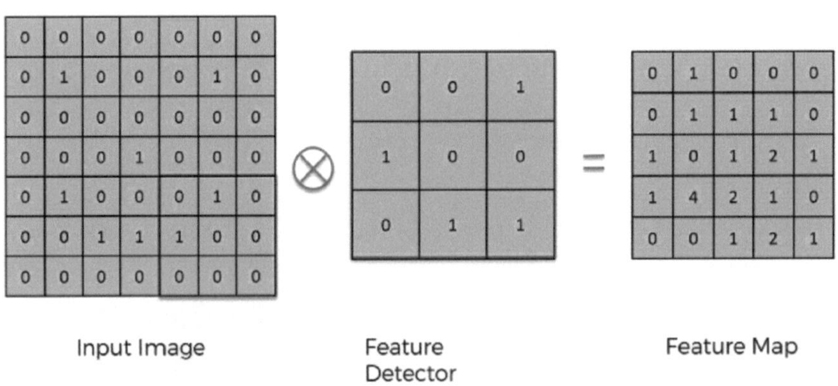


**Figure 4.**
*Display of convolution function of convolutional layer.*

#### *1.4.3 Separable convolution*

In this type of convolution, the kernels are separated into two smaller kernels. A 2D kernel will be broken into row matrix and column matrix. This reduces the number of computations. Despite reduction in the computational cost, this type of convolution is not preferred generally in many deep learning algorithms as it provides only suboptimal results.

#### *1.4.4 Flattened convolution*

In this type of convolution, a 1D kernel is used to traverse over the 3D input volume to produce the output feature maps. This greatly reduces the number of learnable parameters and results in less computational cost.

### 1.5 Pooling layer

Pooling layer is an important layer in CNN. Pooling is used for down sampling. It is mainly involved in reducing the spatial dimension of the feature maps. This considerably reduces the number of parameters required for training the network. By reducing the trainable parameters, it minimizes the computations required. Pooling layer becomes essential in networks with a greater number of layers. Applying a $2 \times 2$ pooling filter reduces the size of the feature map to half of its original size. The figure demonstrates the effect of $2 \times 2$ pooling filter applied on $224 \times 224 \times 64$ feature map that reduces the size of the o/p feature map to $112 \times 112 \times 64$. However the depth of the feature map remains unaltered (**Figure 5**) [6].

There are two types of pooling filters: maximum pooling (max pooling) and average pooling (avg pooling). A $2 \times 2$ max pooling filter selects a $2 \times 2$ region from the input image and passes the maximum value to the next stage. Average pooling filter passes the average value of the selected region to the next stage (**Figure 6**).

### 1.6 Activation functions

The input for the any CNN network is an image, and the output is the class score. This establishes a nonlinear relation between the input and output data. Activation functions are the nonlinear transformation inserted in between the layers. Firing of neurons depends on the results of the activation functions. The important characteristic of an activation function is that its derivative should exist (**Figure 7**).

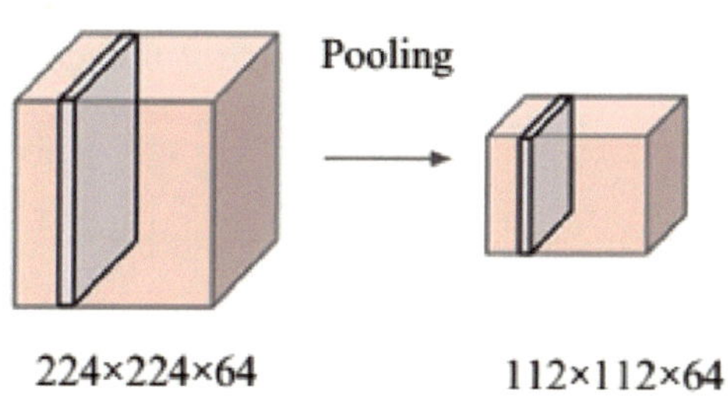


**Figure 5.**
*Effect of pooling layer with $2 \times 2$ filter.*

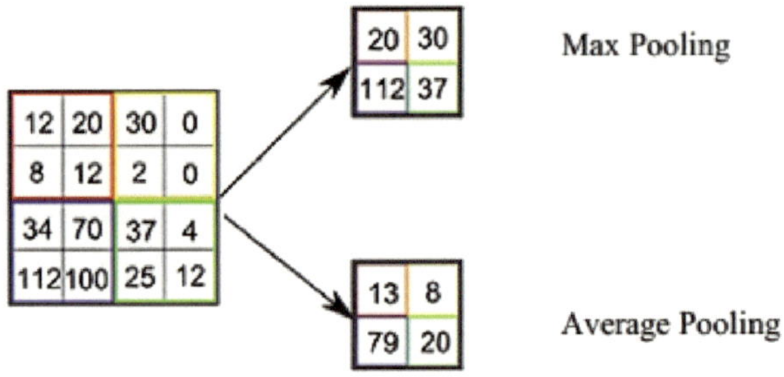


**Figure 6.**
*Application of* $2 \times 2$ *max pooling filter and average pooling filter on a* $4 \times 4$ *feature map.*

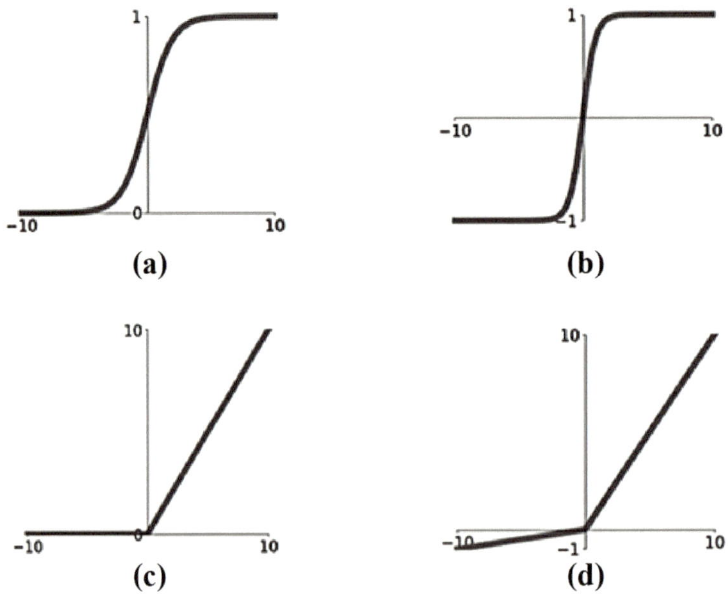


**Figure 7.**
*Activation functions: (a) sigmoid function, (b) tanh function, (c) ReLU function, and (d) leaky ReLU function.*

The activation functions are shown in the figure in which the y-axis represents the function f(a) and x-axis denotes the value a. The following functions are the most prevalent activation functions used in neural network.

## 1.7 Sigmoid function

Sigmoid or logistic function is a function that ranges between 0 and 1. It is computed as given in the equation. Slow convergence of the function makes it less popular.

$$f(a) = \frac{1}{1 + e^{-a}} \tag{5}$$

## 1.8 Hyperbolic tangent function

The hyperbolic tangent function ranges between −1 and 1. It overcomes the disadvantage of sigmoid function of having the zero centered function. It makes the optimization easier and it is preferred over sigmoid function.

$$f(a) = \tanh(a) = (e^{a} - e^{-a})/(e^{a} + e^{-a}) \tag{6}$$

### 1.9 ReLU: rectified linear units

The function results in values in the range $[0, \infty)$. It is the most commonly used activation function. The negative values of the input matrix are removed completely. A simple thresholding can be used for implementing this activation. This function avoids the simultaneous activation of all neurons, thereby reducing the number of computations. Convergence of this function is also faster than the sigmoid and tangent functions. The disadvantages of this function are that it cannot update all the weights during back propagation when the gradient is zero and it results in more number of dead neurons when used with high learning rate.

$$f(a) = \begin{cases} 0 \text{ for } a < 0 \\ a \text{ for } a \geq 0 \end{cases} \tag{7}$$

### 1.10 Leaky ReLU: rectified linear units

The function results in values in the range $(-\infty,\infty)$. Instead of removing negative values, it provides a negative slope for $a<0$. The values of slope can be very small in the range $\delta = 0.01, 0.1$, etc. This eliminates the problem of dying neurons as in the case of ReLU.

$$f(a) = \begin{cases} \delta a \text{ for } a < 0 \\ a \text{ for } a \geq 0 \end{cases} \tag{8}$$

### 1.11 Normalization layer

The stability of the network can be improved by introducing batch normalization layers in between the regular layers. In this layer the mean and standard deviation of the values in previous layer are computed. Then the mean value is subtracted from each of the input value and divided by the standard deviation. The batch normalization gives a choice to network from getting rid of dropouts and enables the network to have a higher learning rate. It introduces two more learnable parameters p1 and p2 to the network. The output y after the batch normalization of a small batch of the input data $x_i = x_1, x_2, x_3, .x_m$ is given by the equation:

$$y = BN_{p_1p_2}(x_i) = p_1\hat{x}_i + p_2 \tag{9}$$

where batch mean $\mu_B = \frac{1}{m}\sum_{i=1}^{m} x_i$ and variance is given as $\sigma_B^2 = \frac{1}{m}\sum_{i=1}^{m}(x_i - \mu_B)$.

### 1.12 Fully connected layer

The name fully connected layer is coined since each neuron in the layer is connected to every neuron in the previous layer as displayed in **Figure 8**. Generally, two to three fully connected layers are appended as the final layers in every CNN. Activations of fully connected layer follow affine transformation, which involves matrix multiplication and addition of bias values. This fully connected layer is equivalent to the classifiers, whereas the previous layers are responsible for feature extraction [7].

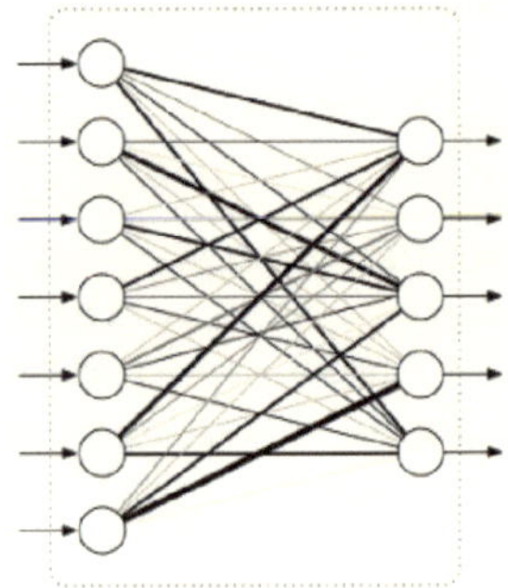


**Figure 8.**
*Fully connected layer.*

## 1.13 Optimization methods

A CNN classifier is involved in calculating the output class score (Y) from the input image with a set of predictors (X). Optimization algorithms aid the network to curtail the error function by updating the parameters in the right direction. Effective training of the network is greatly influenced by the choice of correct network parameters. Optimum values of such parameters can be achieved by adopting suitable optimization algorithms. Few commonly used optimization algorithms are described below.

### *1.13.1 Stochastic gradient descent (SGD)*

A matrix formed with first-order derivatives is called Jacobian matrix, and it is used to represent a gradient function. SGD is an iterative method that is used with batch processing. The network weights are updated with the average value of the gradients of a batch. If L(w) is the loss function or error function, the objective of the SGD is to identify the parameter Ï‰ that minimizes L(w). The SGD algorithm performs the iterations using equation to compute the parameter value.

$$w^{k+1} = w^k - \eta\delta L(w) = w^k - \eta\sum_{i=1}^{n} L(w) \tag{10}$$

where $\eta$ is the learning rate

### *1.13.2 SGD with momentum*

In this algorithm the update for the parameter Ï‰ is computed as a linear combination of the previous update and the gradient. This algorithm prevents oscillation by traveling in the same direction.

$$\delta w = \alpha\delta w - \eta\delta L_i(w) \tag{11}$$

$$w^{k+1} = w^k + \delta w \tag{12}$$

$$w^{k+1} = w^k + \delta w = \alpha\delta w - \eta\delta L_i(w) \tag{13}$$

where $\eta$ is the learning rate and $\alpha$ is the momentum term usually 0.9 or a similar value.

#### *1.13.3 Nesterov's accelerated gradient (NAG)*

NAG utilizes the momentum to predict the future values of the parameter. Thus, the momentum $w - \alpha v_{t-1}$ is used to estimate the next possible position of the parameter. This enables the algorithm to converge more rapidly.

#### *1.13.4 AdaGrad*

AdaGrad algorithm is an adaptive gradient algorithm in which a flexible learning rate is adopted. Larger learning rate is used for features that occur very rarely, and smaller learning rate is used for frequently occurring features. The learning of network stops at a point when the learning rate shrinks to zero. This is a drawback for this method.

#### *1.13.5 RMSProp*

RMSProp also adopts the idea of using a variable learning rate. The learning rate is decided based on the signs of the previous two gradients. If the previous two gradients are of the same sign, it increases the step size in the order of 1.2 times the previous learning rate. Otherwise the step size is decreased in the order of 0.5 times the previous learning rate. The learning rate is always limited between one millionth and 50.

#### *1.13.6 Adam optimizer*

Adam optimizer is the name derived from adaptive momentum estimation. Adam optimizer attracts with its advantages like simple implementation procedure, minimum memory requirements, and efficient computation. The algorithm is best suited for very big data size with less number of tuning and problems with sparse gradients. Adam optimizer retains the benefits of both RMSProp and AdaGrad optimizers. Apart from the first moments of gradients, the mean of the second moments of gradients were also used in Adam optimizer.

### 1.14 Softmax layer

The final layer of CNN architecture is mostly the softmax layer. Normalized exponential function is called the softmax function. This layer converts an N number of input vectors into N probabilities. This layer is essential as it normalizes the input vector of any value (negative values, positive values greater than one, etc.). The number of nodes in the softmax layer is equal to the number of output classes. The softmax function for the input vector $z_i = z_1, z_2, z_3, z_N$ is converted in to probabilities $P(z)_i$ using the given equation:

$$P(z_i) = \frac{e^{zi}}{\sum_{j=1}^{N} e^{zi}} \tag{14}$$

## 2. LeNET architecture and its applications

Optical character recognition is one of the major research problems in real-time applications, and it is used to recognize all the characters in an image. As English is a universal language, character recognition in English is a challenging task. Deep

| Layer | Feature map | Size | Kernel size | Stride | Activation |
|---|---|---|---|---|---|
| Input layer | 1 | 32 × 32 | — | — | — |
| Convolution layer 1 | 6 | 28 × 28 | 5 × 5 | 1 | tanh |
| Maximum pooling layer | 6 | 14 × 14 | 2 × 2 | 2 | tanh |
| Convolution layer 2 | 16 | 10 × 10 | 5 × 5 | 1 | tanh |
| Maximum pooling layer 2 | 16 | 5 × 5 | 2 × 2 | 1 | tanh |
| Convolution layer 3 | 120 | 1 × 1 | 5 × 5 | 1 | tanh |
| Fully connected layer | — | 84 | — | — | tanh |
| Output layer (FC 2) | — | 10 | — | — | Softmax |

**Table 1.**
*Summary of LeNET architecture.*

learning approach is one of the solutions for the recognition of optical characters. The aim of this research work is to perform character recognition using convolutional neural network with LeNET architecture. Optical character images have been binned in each class (10 classes) to form 2495 samples of training images and 1069 test images. Each of the images fed for training/testing has a size of 32 × 32 such that the convolution layer is enabled with a filter size of 5 × 5. Thus the output of the convolution layer is 28 × 28 along with six feature maps [24, 25]. The second (maximum pooling) layer gets a maximum of 2 × 2 value from the 28 × 28 image. The maximum pooling layer [21], [22] thus gives an output size of 14 × 14 with six feature maps. The convolution layer forms the third layer which performs the convolution operation of input with filter. The output of this layer is 10 × 10 with 16 feature maps for a filter size of 5 × 5. The last layer in the network is a convolution layer with filter size of 5 × 5. The output of the last layer is fed to fully connected layers 1 and 2. The fully connected layer 2 forms the output layer with a total of 10 outputs (**Table 1**).

## 3. CNN applications

### 3.1 AlexNET architecture and its applications

Image classification is one of the major research problems in real-time applications, and it is used to recognize all the objects in an image. Deep learning approach is one of the solutions for the recognition and identification of different images. The aim of this research work is to perform character recognition using convolutional neural network with AlexNET architecture. For analysis of AlexNET architecture with the created database, an input image is fixed with a size of [227 × 227]. The first layer in AlexNET is the convolution layer 1 with an input size of [227 × 227]. This layer convolves the input with a filter size of [11 × 11] and provides an output with 96 feature maps with a size of [55 × 55] (stride 4) [8].

Maximum pooling layer forms the second layer, which subsamples the output of the first layer with a pooling size of [3 × 3] and gives an output size of [55 × 55] (96 feature maps). The third layer convolves the output of the second layer with a filter size of [5 × 5]. The output of the third layer is once again pooled with a filter (size 27 × 27) to get a feature map vector of size 256. The maximum size of filter utilized in the next layer is [3 × 3] for an image input of [27 × 27], thereby providing an output of 256 feature maps. Another convolution layer forms the sixth layer which convolves with the fifth layer to produce a feature vector size of 256.

| Layer | Feature map | Size | Kernel size | Stride | Activation |
|---|---|---|---|---|---|
| Input layer | 1 | 227 × 227 | — | — | — |
| Convolution layer 1 | 96 | 55 × 55 | 11 × 11 | 4 | ReLU |
| Maximum pooling layer | 96 | 27 × 27 | 3 × 3 | 2 | ReLU |
| Convolution layer 2 | 256 | 27 × 27 | 5 × 5 | 1 | ReLU |
| Maximum pooling layer 2 | 256 | 13 × 13 | 3 × 3 | 2 | ReLU |
| Convolution layer 3 | 384 | 13 × 13 | 3 × 3 | 1 | ReLU |
| Convolution layer 4 | 384 | 13 × 13 | 3 × 3 | 1 | ReLU |
| Convolution layer 5 | 256 | 13 × 13 | 3 × 3 | 1 | ReLU |
| Max pooling | 256 | 6 × 6 | 3 × 3 | 2 | ReLU |
| Fully connected layer 1 | — | 9216 | — | — | ReLU |
| Fully connected layer 2 | — | 4096 | — | — | ReLU |
| Fully connected layer 3 | — | 4096 | — | — | ReLU |
| Output layer (FC -4) | — | 10 | — | — | Softmax |

**Table 2.**
*AlexNET architecture summary for palm leaf characters.*

**Table 2** gives a summary of the AlexNET architecture that has been created during this experiment.

### 3.2 FaceNet architecture and its applications

Face detection is the process of detecting a face in an image or a video. Face recognition is the process of detecting face in an image and then using algorithms to identify who the face belongs to. Face recognition is thus a form of person identification. Initially, features are extracted from the image for training the machine learning classifier to identify faces in the image. Not only are these systems not subjective, but also they are also automatic—no hand labeling of facial features is required. Since for face recognition, we need to detect a face from the image or video, we can think of face recognition as a two-phase stage: Stage 1, detecting the presence of faces in the image or video stream using methods such as Haar cascades, HOG+Linear SVM, deep learning, or any other algorithm that can localize faces, and Stage 2, taking each of these faces detected during the localization phase and

| Type | Patch size/stride | Output size |
|---|---|---|
| Convolution | 34 × 34/2 | 112 × 112 × 64 |
| Max pool | 3 × 3/2 | 56 × 56 × 64 |
| Convolution | 3 × 3/1 | 56 × 56 × 192 |
| Max pool | 3 × 3/2 | 28 × 28 × 192 |
| Inception(3a) | — | 28 × 28 × 256 |
| Inception(3b) | — | 28 × 28 × 480 |
| Max pool | 3 × 3/2 | 14 × 14 × 480 |
| Inception(4a) | — | 14 × 14 × 512 |
| Inception(4b) | — | 14 × 14 × 512 |
| Inception(4c) | — | 14 × 14 × 512 |

| Type | Patch size/stride | Output size |
|---|---|---|
| Inception(4d) | — | 14 × 14 × 528 |
| Inception(4e) | — | 14 × 14 × 832 |
| Max pool | 3 × 3/2 | 7 × 7 × 832 |
| Inception(5a) | — | 7 × 7 × 832 |
| Inception(5b) | — | 7 × 7 × 1024 |
| Avg pool | 7 × 7/1 | 1 × 1 × 1024 |
| Dropout (40%) | — | 1 × 1 × 1024 |
| Linear | — | 1 × 1 × 7 |
| Softmax | — | 1 × 1 × 7 |

**Table 3.**
*FaceNet architecture.*

learning whom the face belongs to; this is where you assign a name to a face.
Face detection is the techniques to finding all the faces in an image or videos [9]. Face recognition is the next step after face detection. In face recognition you identify which face belongs to which person using an existing (Pre-Trained Image) image in the repository. Face analysis is a process of extracting facial features like age, complexion, etc., from the image after recognizing a face in it. Generally, OpenCV provides three different methods for face recognitions like eigenfaces [8], local binary pattern histograms, and fisherfaces. But, nowadays, deep learning using FaceNet is a very popular algorithm used in many applications, which is shown in **Table 3**.

### 3.3 Object detection and its applications

The environmental problems and its treatment go back to the fifteenth century. A wide variety of road types such as intersections and highways pose a real challenge to the computer vision algorithms. Hence, there is a need of efficient algorithm to detect the accident on road and also evaluate the severity of the incident. In this paper, an accident detection approach using ResNet architecture has been presented with specific focus on road accident. The convolutional neural network used in this paper has utilized around 50 layers, viz., convolution layer, pooling layer, activation layer, fully connected layer, and softmax classifier and inspection layer. The paper has also created a database of accident video set by utilizing the video images of accident. The recognition of ResNet 50 classifier has been found to be 98.1%. The prediction rate is found to be higher due to the large quantum of features extracted in each of the CNN layers. There is no space for waste. Our landfill sites are filling up fast; by 2010, almost all landfills in the UK will be full. Financial expenditure in the economy is reduced. Creating products from raw materials costs much more than if they were made from recycled products. Natural resources should be preserved for future generations. Recycling reduces the need for raw materials; it also uses less energy, therefore preserving natural resources for the future garbage sorter robot, a device which uses image processing combined with robotics to collect and sort the available garbage using highly sensitive sensors and a smartphone camera. This would be very useful in cleaning the environment and for effective recycling without wasting man power. Garbage is an important cause of many diseases. Recycling must be made more effective. Usage of man power must be considerably reduced which is shown in **Figure 9**. This being our

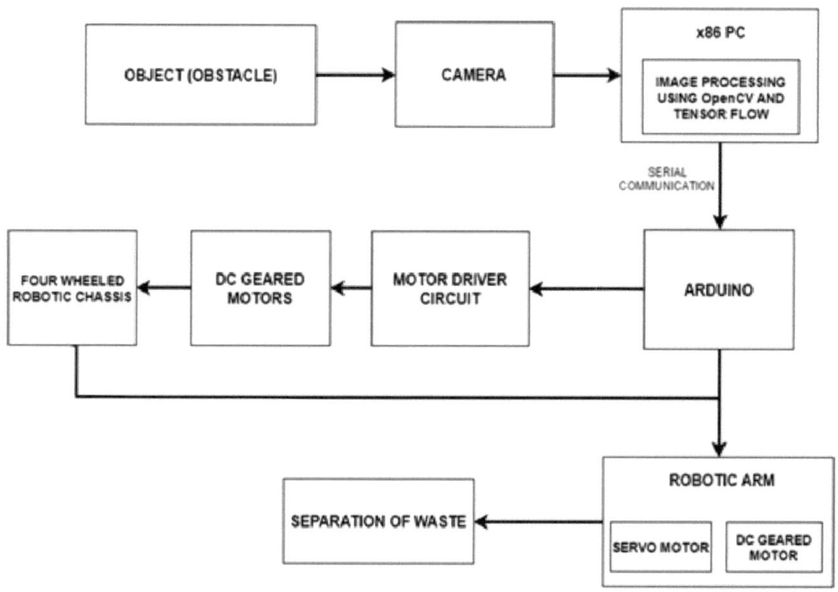


**Figure 9.**
*Object detection system.*

motivation, we could take some naturally available parameters that come out of the garbage and make a device that could read them and detects the changes in them. Thus by doing so, the negative effects due to waste could be reduced. The above gives us a clear picture of the hardware and software used accordingly. The components used along with their usage are explained below:

### *3.3.1 Camera*

A webcam which is around 5 MP is used for capturing the images, and the captured images are fed to the ×86 PC for processing them.

### *3.3.2 ×86 PC*

The required image processing software, i.e., OpenCV and TensorFlow, is installed and used for processing the captured images. Here the language used is Python. Similarly the Arduino IDE is used to program the Arduino board. Arduino board is used for the movement of wheels and robotic arm which is shown in **Figure 10**.

### *3.3.3 Robotic base*

The robotic base consists of an acrylic sheet which has two pair of wheels whose movement is done via a DC geared motor. The base also has a robotic arm.

### *3.3.4 Robotic arm*

The arm consists of a gripper which is connected to an actuator that moves up and down. There is another actuator connected to the previous one which moves back and forth; this movement is done using DC geared motors. The below given is the flow diagram which shows the work flow of the system.

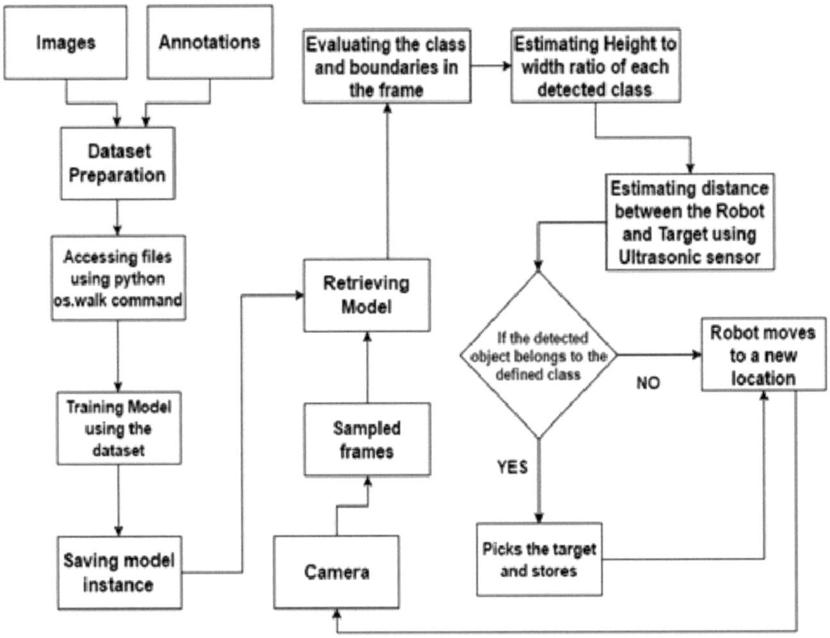


**Figure 10.**
*Flow diagram of the system.*

Classification of object is done using image processing and TensorFlow. Camera is used to capture the image of the object present on the surface which is then matched with the pre-trained dataset. Here the object detection is achieved using YOLO object detection, and the detected object is directed to its respective class. There are several modes that are made available such as plastic, glass, metal, degradable, and nondegradable. Based on the information of the selected mode, either the object is picked up or ignored. The object is picked up using a robotic arm which is controlled using the pre-programmed Arduino board. Once the desired object is picked up in accordance with the selected mode, object classification is completed successfully. Advances are done in proposed method in comparison with the existing method:

- Some areas in the design of the robot in existing method have limitations; these limitations may be overcome by improving upon the design.

- Thus we have made some improvements in the design of the robot by making it a bit smaller, adding a webcam which is weightless and making the wheel base strong and stable.

- We have also removed the bins available on top of the robot as the previous one only classifies between two classes but our robot classifies almost between more than three classes.

- The above robot mentioned in the existing method uses manual pickup of the classified waste, but we have added a robotic arm gripper which automatically picks up the classified object.

- Also the Recyclebot is controlled manually over a Zigbee powered joystick application, but our robot is fully automated and uses Arduino to achieve it.

- On the software side, we have made some major improvements like we have used OpenCV and TensorFlow for processing the obtained image.
- Also we have used an algorithm called YOLO for object detection and classification which provides better accuracy when compared with the previous method.
- Since the robot moves by itself, the number of hardware and software used is reduced which means there is no need for Zigbee and the joystick application anymore.
- When it comes to recognizing objects, the Recyclebot only recognizes a small set of images which is around 1000 to 2000.
- But our robot could almost recognize more than 5000 images which prove to be a larger image set than the previous one.
- Also the speed of processing is improved. The accuracy of our robot is far better for an automated robot.
- Thus this shows that our robot is much better than its predecessor.

#### *3.3.5 Skin cancer classification using convolutional neural network*

Skin diseases are becoming the most common health issues worldwide. In this paper we propose a method that detects four types of skin disease using computer vision [10, 11]. The proposed approach involves convolutional neural networks with specific focus on skin disease. The convolutional neural network used in this paper has utilized around 11 layers, namely, convolution layer, pooling layer, activation layer, fully connected layer, and softmax classifier. Images from the DermNet database are used for validating the architecture. The database comprises all types of skin diseases out of which we have considered four different types of skin diseases like acne, keratosis, eczema herpeticum, and urticaria with each class containing around 30–60 different samples. The challenges in automating the process includes the variation of skin tones, location of the disease, specifications of the image acquisition system, etc.

**Author details**

Anand Raju*† and Shanthi Thirunavukkarasu†
Department of Electronics and Communication Engineering, Sona College of Technology, Salem, Tamil Nadu, India

*Address all correspondence to: anandvimal1@gmail.com

† These authors contributed equally.

## References

[1] Sainath TN, Mohamed AR, Kingsbury B, Ramabhadran B. Deep convolutional neural networks for LVCSR. In: 2013 IEEE International Conference on Acoustics, Speech and Signal Processing. IEEE; 2013. pp. 8614-8618. Available from: http://www.cs.toronto.edu/~asamir/papers/icassp13_cnn.pdf

[2] Simard PY, Steinkraus D, Platt JC. Best practices for convolutional neural networks. In: International Conference on Document Analysis and Recognition (ICDAR). 2003. p. 958

[3] Lawrence S, Giles CL, Tsoi AC, Back AD. Face recognition: A convolutional neural-network approach. IEEE Transactions on Neural Networks. 1997; **8**(1):98-113

[4] Liu L, Shen C, van den Hengel A. The treasure beneath convolutional layers: Cross-convolutional-layer pooling for image classification. In: Proceedings of the IEEE Conference on Computer Vision and Pattern Recognition. 2015. pp. 4749-4757

[5] Krizhevsky A, Sutskever I, Hinton GE. ImageNet classification with deep convolutional neural networks. In: Advances in Neural Information Processing Systems. 2012. pp. 1097-1105

[6] Graham B. Fractional max-pooling. 2014. arXiv preprint arXiv:1412.607

[7] Anand R, Shanthi T, Sabeenian RS, Veni S. Real time noisy dataset implementation of optical character identification using CNN. International Journal of Intelligent Enterprise. 2020;**7** (1-3):67-80

[8] Anand R, Kalkeseetharaman PK, Naveen Kumar S. Automatic facial expressions and identification of different face reactions using convolutional neural network

[9] Anand R, Shanthi T, Nithish MS, Lakshman S. Face recognition and classification using GoogleNET architecture. In: Soft Computing for Problem Solving. Singapore: Springer; 2020. pp. 261-269

[10] Sabeenian RS, Paramasivam ME, Anand R, Dinesh PM. Palm-leaf manuscript character recognition and classification using convolutional neural networks. In: Computing and Network Sustainability. Singapore: Springer; 2019. pp. 397-404

[11] Shanthi T, Sabeenian RS, Anand R. Automatic diagnosis of skin diseases using convolution neural network. Microprocessors and Microsystems. 2020;**76**:103074

Chapter 4

# Estimation for Motion in Tracking and Detection Objects with Kalman Filter

*Afef Salhi, Fahmi Ghozzi and Ahmed Fakhfakh*

## Abstract

The Kalman filter has long been regarded as the optimal solution to many applications in computer vision for example the tracking objects, prediction and correction tasks. Its use in the analysis of visual motion has been documented frequently, we can use in computer vision and open cv in different applications in reality for example robotics, military image and video, medical applications, security in public and privacy society, etc. In this paper, we investigate the implementation of a Matlab code for a Kalman Filter using three algorithm for tracking and detection objects in video sequences (block-matching (Motion Estimation) and Camshift Meanshift (localization, detection and tracking object)). The Kalman filter is presented in three steps: prediction, estimation (correction) and update. The first step is a prediction for the parameters of the tracking and detection objects. The second step is a correction and estimation of the prediction parameters. The important application in Kalman filter is the localization and tracking mono-objects and multi-objects are given in results. This works presents the extension of an integrated modeling and simulation tool for the tracking and detection objects in computer vision described at different models of algorithms in implementation systems.

**Keywords:** Kalman filter, tracking objects, detection objects, localization, video and image processing, computer vision, embedded system

## 1. Introduction

The computer vision, from the technological evolution point of view, is the most useful in our days. It is a discipline at the border of computer science, mathematics, physics, neuroscience and various other disciplines, which aims to initiate the specific issues of image and video analysis from *2D* and *3D* environment, and to implement a simple object tracking application. This phenomenon provokes a spectacular development of applications in various fields in many sectors of activity: imaging systems, robotics, surveillance systems, identification of interest (automatic annotation and retrieval of video from databases multimedia data), indexing and augmented reality, HMI interaction (gesture and gaze recognition for data entry on computers), etc. These systems are most used in airports, metros, prisons, banks or nuclear power plants, intelligent transport systems, the analytical approach for medical applications, military imagery with the target weapon, applications security and computer-controlled automatic surveillance (scene surveillance, object tracking and behavior analysis, swimming systems for swimming pool surveillance, to prevent accidents and drowning victims), video conference, driving assistance (reversing radar, speed limiter or cruise control), pedestrian tracking (counting and pedestrian tracking systems using aerial cameras), biometric systems (fingerprints and recognition biometric facial), etc. Such an application uses computer vision techniques: object detection, classification of moving objects and tracking of objects, etc. The main objective is to locate a known object in the image in order to follow it up such as faces, people, hand gestures, cars, etc. The current trend is to lighten the tasks performed by humans by integrating intelligence into these systems. So in computer vision, the tracking of moving objects in a known or unknown environment is commonly studied since the year 1970. Monitoring can be a tool to give visual autonomy to robots. In this case, visual perception is a prerequisite for action and requires learning to establish links between the causes and the actions to be produced in response. Tracking objects can also automate repetitive tasks. Such a monitoring system must be robust to the following real constraints: variations in the lighting of the sequence, change in the pose of the object (front view, profile view), change in scale (change in size of the object), change of appearance, simultaneous movement of the camera and the object, partial and total occlusions, or even the kinematics (for example the space-time constraints) and low processing time (20 images/s). Our aim is to classify these methods efficiently in order to highlight the advantages and disadvantages of each method. This will allow us, later, to choose the most robust algorithm for an object tracking system. The object tracking system uses the method of tracking the region of interest of the object in a video sequence. Several points will be discussed, such as the pre-treatment methods, the change of the object and its movement, the change of appearance, the change of scale, and change of illumination. Then we will compare the tracking results for the different video sequences analyzed and show the performance of the implemented algorithm.

Tracking corresponds to the estimation of the location of the object in each of the images in a video sequence, the camera and/or the object (face, man, hand, animal, etc.) being able to be simultaneously in motion. The localization process is based on the recognition of the object of interest from a set of visual characteristics (color, shape, speed, etc.). Specifically, the purpose of an object tracking method is to estimate, in each image of the sequence, the functions that are used in tracking the object or objects present in the field of vision of the camera such as motion, color, corners, outline, shape, and object view. In object tracking, the class, appearance, scale, and/or location of the tracking region are predicted based on the forward images and on the underlying model for state transitions. The state of the object is generally represented by its location and its speed.

There are then three main stages in the analysis of the video sequence, the first stage consists in carrying out the detection of moving objects. Then the step of tracking these objects from one image to another and finally, we analyze the tracks of objects to recognize their behavior. Many different techniques for tracking objects have been proposed. The detection events and detection moving objects in complex scenes is difficult to analyze due to camera noise and changing lighting conditions. Each limitation must be overcome in order to avoid failure of the tracking algorithm. In an object tracking algorithm, there are generally four steps: detection, location, association, and trajectory estimation [1–3]. The algorithms are composed by three important modules: block matching and meanshift, camshift, Kalman filter. The Kalman filter is used in a wide range of technological fields. It is a major theme in automation and frame and signal processing. The Kalman filter

"KF" is a set of mathematical equations which provide an efficient (recursive) computation of the means for estimation the state of a process. The KF is very powerful in several aspects: it supports estimates of past, present and even future states and it can do so even when the exact nature of the modeled system tracking and detection objects. The Kalman filter is a corrective predictor filter. In the tracking system objects, this filter looks at an object as it moves, that is, it takes information on the state of the object at the precise moment. Then, it uses this information to predict where the object is in the next frame. For this, it takes as input a measurement vector (position in x, in y, width and height of the object). In the tracking process, this filter looks at an object as it moves, that is it takes information on the state of the object at the precise moment. In the case of tracking an object in motion, the Kalman filter allows us to estimate the states of motion of the object (and therefore predetermine the areas of motion in the following frames with using the combination for three algorithms (block-matching, Camshift and Kalman Filter)) and thus adds robustness tracking objects. Many authors have studied the Kalman filter in object tracking [1, 2]. In this work, we optimized many criteria in image and video processing application. For example, we can site: time execution, quality and performance in the image and video processing, artifact and noise in a frame, etc., the data flow for Kalman Filter is presented in **Figure 1**.

## 2. Different methods of modeling an object

In a follow-up scenario, an object can be defined as anything that is of interest for further analysis. For example people walking on a road, boats on the sea, fish inside an aquarium, airplanes in the air, cars on the road, a motion hand or face, motion for different objects and multi-objects, etc. It is a collection of objects that may be important to track in a specific area or environment. The implementation of an object tracking system involves designing two main parts, the object representation and the object location. The localization step is based on the representation model of the object and its location in the previous frame. The representation consists in associating with the object followed characteristics of shape and/or appearance allowing to recognize it in successive frames. In recent studies, representations by shape and appearances are classified into three families such as representation by point clouds, representation by bounding boxes (representation by geometric shapes) and representation by silhouettes. In what follows, we will describe these methods as illustrated in **Figure 2** [4–6].

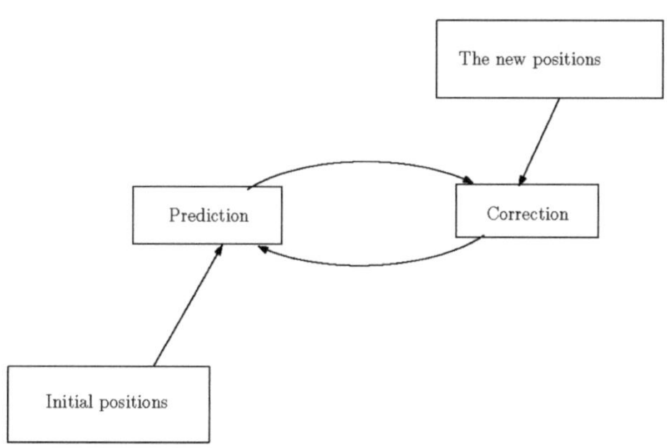


**Figure 1.**
*The data flow for a Kalman filter.*

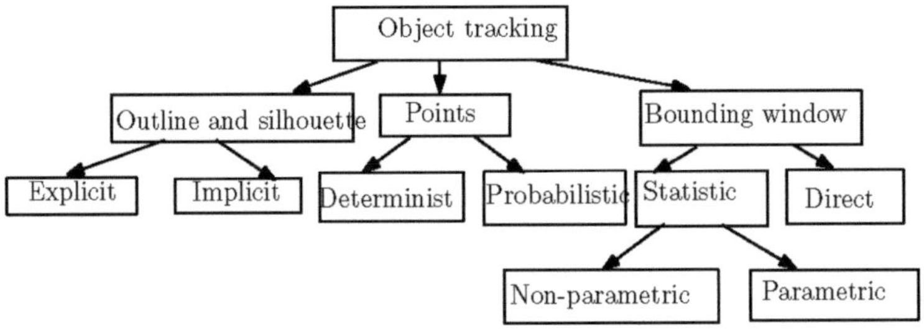


**Figure 2.**
*Object tracking methods.*

1. What representation is appropriate for tracking objects?

2. What algorithm should be used?

3. How is the movement, appearance and shape of the object modeled?

Tracking of events and detection motion objects in complex scenes is difficult to analyze due to camera noise and changing lighting conditions. Each limitation must be overcome in order to avoid failure of the tracking algorithm. In an object tracking algorithm, there are generally four steps: object detection, location, association, and trajectory estimation. We will be interested in this master's work [1, 2, 7] to study the different methods of representing objects in a video sequence.

### 2.1 Camshift algorithm

Camshift is an algorithm for tracking objects in real time (people, vehicles). It is based on the colors developed in the video sequence. Camshift is based on the average displacement algorithm (Meanshift). The calculation module is based on iterations to reach convergence. Camshift take the HSV color space as a model with the color tone component (*Hue*). This component is designed to calculate the probability of the histogram of each image of the analyzed sequence. The size of the original window was just large enough to fit most of the object inside of it. The Camshift algorithm adjusts the size of the search window according to the movement of the object analyzed with constant tint. Whereas, for a quick movement, the follow-up can fall into the analysis of another object in the sequence. For this reason, we choose at the beginning of the algorithm a threshold of color hue of the object to ensure correct tracking. Once the mean displacement module converges, the center of gravity and the zero order moment are calculated. Then we calculate the new size, width and length of the search window. Then, the window is centered around center of gravity and the calculation of the next image is started. Next, we calculate the Camshift parameters such as the secondary moments, the orientation, the width and the length of the window around the object's center of gravity. **Figure 3** shows the flowchart of the Camshift algorithm for object tracking.

#### *2.1.1 Calculation of Camshift parameters*

With each iteration of the Camshift algorithm, the object search window will be resized. To search for the new size, the search window obtained by the average displacement algorithm is slightly enlarged to include the object. Then, the parameters of the window must be adapted such as the width, the length and the center of gravity. The term $M00/255$ is the normalized area since the zero order moment is calculated

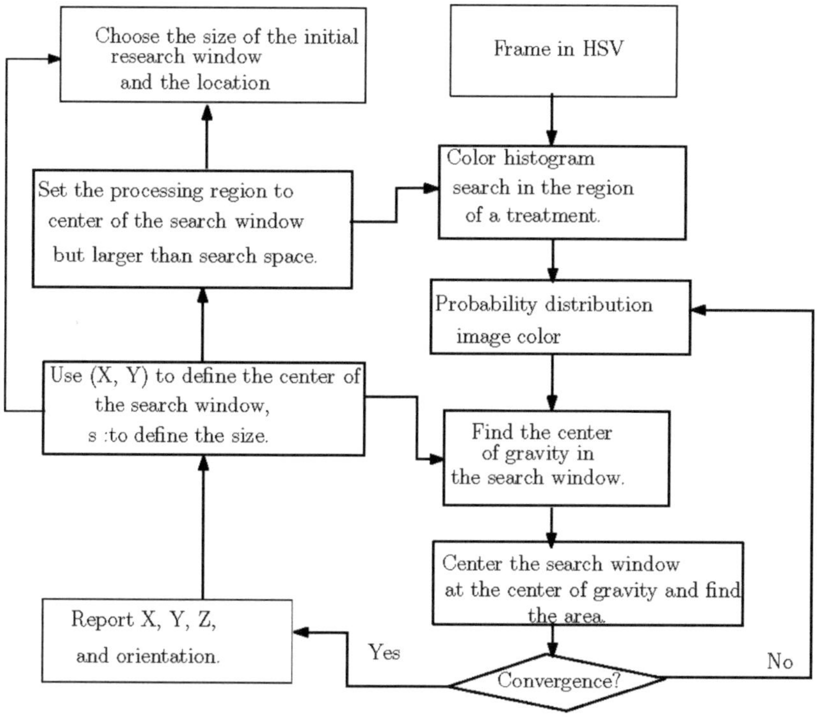


**Figure 3.**
*Flowchart of the Camshift algorithm for object tracking.*

from the probability distribution of the image which can have values from 0 to 255. The search window adaptation parameters are computed in Eqs. (1), (2) and (3) [1, 7].

$$s = 2 \times \sqrt{\frac{M00}{255}} \tag{1}$$

$$W = [s \quad 1.2 \times s] \tag{2}$$

$$Lc = [(xc - (W(1)/2)) \quad (yc - (W(2)/2))] \tag{3}$$

We calculate the secondary moments using the Eqs. (4), (5) and (6), orientation is calculated by Eq. (7) and the length and width of the object search window (11) and (12).

$$M20 = \sum_x \sum_y x^2 \times I(x,y) \tag{4}$$

$$M02 = \sum_x \sum_y y^2 \times I(x,y) \tag{5}$$

$$M11 = \sum_x \sum_y x \times y \times I(x,y) \tag{6}$$

$$\theta = \frac{\arctan \dfrac{2 \times \left(\dfrac{M11}{M00} - xc \times yc\right)}{\left(\dfrac{M20}{M00} - x_c^2\right) - \left(\dfrac{M02}{M00} - y_c^2\right)}}{2} \tag{7}$$

$$\theta = \frac{\arctan \dfrac{2 \times b}{a - c}}{2}$$

$$a = \frac{M20}{M00} - x_c^2 \tag{8}$$

$$b = 2 \times \left( \frac{M11}{M00} - x_c \times y_c \right) \tag{9}$$

$$c = 2 \times \left( \frac{M02}{M00} - y_c^2 \right) \tag{10}$$

$$l = \sqrt{\frac{(a + c) + \sqrt{b^2 + (a - c)^2}}{2}} \tag{11}$$

$$w = \sqrt{\frac{(a + c) - \sqrt{b^2 + (a - c)^2}}{2}} \tag{12}$$

## 2.2 Kalman filter

### *2.2.1 Definition*

The Kalman filter is a set of mathematical equations which provide an efficient (recursive) computation of the means for estimation the state of a process, so as to minimize the mean of the quadratic error. The filter is very powerful in several aspects: it supports estimates of past, present and even future states and it can do so even when the exact nature of the modeled system is unknown. The filter allows, thanks to its role, to correct and restrict the areas in which we seek movement in the next step. We can see the quadratic error by Eq. (13).

$$E_x = x_{Kalman} - x_{tracking} \tag{13}$$

### *2.2.2 Role of the Kalman filter in the tracking application*

The Kalman filter is used in a wide range of technological. In the tracking and detection process, this filter looks at an object as it moves, that is, it takes information on the state of the object at the precise moment. Then, it uses this information to predict where the object is in the next frame. For this, it takes as input a measurement vector (position in x, in y, width and height of the object). Then it acts on so-called internal parameters (position, speed and acceleration in x and y, as well as the height, the width) to make a prediction and then an estimate of these. Finally, the result is an estimate of the following measurement. In the case of tracking an object in motion, the Kalman filter allows us to estimate the states of motion of the object. Many authors have studied the Kalman filter in object tracking [1, 8, 9], the differences of the present work and the earlier works are the type and the method of objects tracking.

### *2.2.3 Formulation and modeling of the Kalman filter*

The main objective of the Kalman filter is to estimate the vector of states in a discrete time. This process is illustrated by Eq. (14) with stochastic linear differences:

$$x_k = A \times x_{k-1} + w_{k-1} \tag{14}$$

With a measurement vector which has the following form (15):

$$z_k = H \times x_k + v_k \tag{15}$$

$A$ is the transition matrix and $H$ presents the measurement matrix. Random variables $w_k$ and $v_k$ present Gaussian and measurement noise (respectively). They are assumed to be independent (from each other), the covariance of $w_k$ is a matrix $Q$ (16), similarly the covariance of $v_k$ is a matrix "R" (17).

$$p(w_k) \sim N(0, Q) \tag{16}$$

$$p(v_k) \sim N(0, R) \tag{17}$$

The Matrix $A$ models the movement of the object. The movement model used is generally at constant speed or at constant acceleration. Since the movement of objects is not uniform, this type of model is not suitable for describing all movements in general. However, we assume that the object movements that we consider in the third chapter have an adapted dynamic. In our case, we use the motion model with constant acceleration, for tracking a point in two dimensions. The state vector is written:

$$x_k = (x, y, \dot{x}, \dot{y}, \ddot{x}, \ddot{y}) \tag{18}$$

where $(x, y)$ is the position, $(\dot{x}, \dot{y})$ presents the speed and $(\ddot{x}, \ddot{y})$ is the acceleration. Note that in the object tracking application, the noise covariance matrices ($Q$, $R$), the transition matrix $A$ and the measurement matrix are assumed to be constant. The variable parameters over time are the state vector and the measurement vector since the position and size of the object changes during the sequence.

### 2.3 The origins of filter formulation

We define $\hat{x}_k^-$ the estimation of a priori states at the moment $k$ and we give knowledge of the prior process at the moment $k$, and $\hat{x}_k$ the estimate of the state a posteriori at the moment $k$ which is given to the measurement vector $z_k$ (15). We can then define the a priori and a posteriori estimation errors by the following Eqs. (19) and (20).

$$e_k^- = x_k - \hat{x}_k^- \tag{19}$$

$$e_k = x_k - \hat{x}_k \tag{20}$$

The covariance of the a priori estimation error is illustrated by Eq. (21):

$$P_k^- = E\left[e_k^- e_k^{-T}\right] \tag{21}$$

For computing statistics for the Kalman filter, we start with the goal of finding an equation that computes an a posteriori state estimate as a linear combination of an a priori estimate and a weighted difference between an actual measure $zk$ and a measurement prediction $H\hat{x}_k$ as shown in the equation below (22):

$$\hat{x}_k^- = \hat{x}_k^- + K \times \left(z_k - H \times \hat{x}_k^-\right) \tag{22}$$

The difference $\left(z_k - H \times \hat{x}_k^-\right)$ is called measurement innovation. This difference reflects the difference between the predicted measurement and the actual measurement. The $K$ matrix of dimension $n \times m$ in the previous equation (??), is chosen to be the gain or the mixing factor which minimizes the covariance of the posterior error. The gain of the Kalman filter is of the following form (23):

$$K_k = P_k^- \times H^T \times \left(H \times P_k^- \times H^T + R\right)^{-1} \tag{23}$$

The use of a Kalman filter then allows us to estimate the parameters for tracking objects. However, the Kalman filter does not allow the moving element or these parameters to be extracted in the frame. We will first propose a method for detecting the moving object in the frame or video sequence. The fact that we used a robust, reliable and precise tracking algorithm greatly helped us to extract the two measurement and state vectors for the initialization of the Kalman filter (the inputs of the Kalman filter).

## 3. Different function of the Kalman filter

The Kalman filter is an optimal recursive estimator to the linear filtering problem Data. This filter has two necessary modules, a prediction module and a correction or estimation module. The Kalman filter then makes it possible to estimate the position of the object by achieving a compromise between the position observed in the frame and the predicted position. The input parameters of the Kalman filter are respectively, the position of the object in the frame at time "k," the size of the object and the width and length of the object search window which are variable due to the mobility of the object during the sequence. These parameters represent the state vector and the filter measurement vector from Kalman filter. From the works that are studied in the literature [1, 4, 5], we chose the Kalman filter for estimating the tracking parameters. **Figure 4** shows the Kalman filter cycle [8–10]. Generally the estimation of the tracking parameters with a Kalman filter is a process requires the following steps: itemize.

The measure which consists in taking the tracking parameters computed in the Camshift algorithm.

The estimate, which updates the position of the object.

The prediction, which computes the position of the object in the next frame.

The variable parameters of the Kalman filter are the state vector and the measurement vector:

The state vector is composed by the initial position, the width and the length of the search window as well as the center of gravity of the object $(x_c; y_c)$ at time $t_k$. This vector is presented by the following Eq. (24):

$$s_k = \left(x_k; y_k; W_k; L_k; x_c; y_c\right) \tag{24}$$

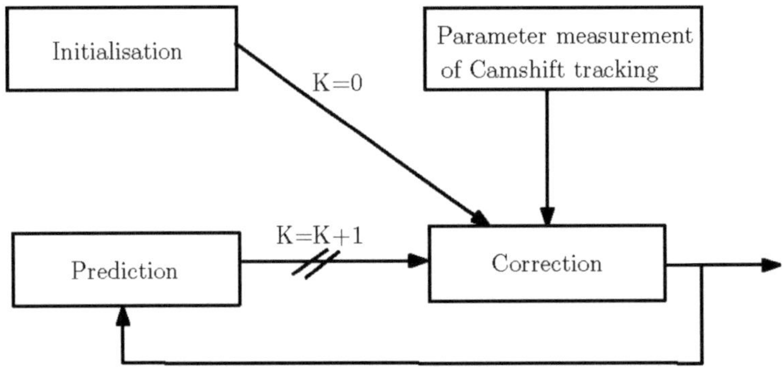


**Figure 4.**
*Kalman filter cycle.*

The measurement vector of the Kalman filter is composed of the initial position, the length and the width of the search window for the object at time $t_k$. This vector is given by the Eq. (25):

$$z_k = (x_k; y_k; W_k; L_k) \tag{25}$$

### 3.1 Process to estimate

The Kalman filter estimates the state "s" of a discrete process, this state is modeled by the linear Eq. (26):

$$s_k = A \times s_{k-1} + w_{k-1} \tag{26}$$

With "A" (27) is the transition matrix, $w_k$ is the process noise and $d_t$ represents the difference between the two instants $k$ and $k-1$ (dt = 1).

$$A = \begin{pmatrix} 1 & 0 & dt & 0 & 0 & 0 \\ 0 & 1 & 0 & dt & 0 & 0 \\ 0 & 0 & 1 & 0 & 0 & 0 \\ 0 & 0 & 0 & 1 & 0 & 0 \\ 0 & 0 & 0 & 0 & 1 & 0 \\ 0 & 0 & 0 & 0 & 0 & 1 \end{pmatrix} \tag{27}$$

- The measurement model is defined by the Eq. (28).

$$z_k = H \times s_k + v_k \tag{28}$$

With $H$ (29) presents the measurement matrix:

$$H = \begin{pmatrix} 1 & 0 & 0 & 0 & 0 & 0 \\ 0 & 1 & 0 & 0 & 0 & 0 \\ 0 & 0 & 1 & 0 & 0 & 0 \\ 0 & 0 & 0 & 1 & 0 & 0 \end{pmatrix} \tag{29}$$

The two vectors $s_k$ and $z_k$ present the state and the measure at the moment $k$, $N$ is the integer vector. Process noise "$w_{k-1}$" and measurement "$v_k$" are assumed to be independent of the state and measurement vectors and to the normal and white distributions which are presented by Eqs. (30) and (31) [9, 10]:

$$p(w) \sim N(0, Q) \tag{30}$$

$$p(v) \sim N(0, R) \tag{31}$$

The noise process is of the following form (32):

$$w_{k-1} = \begin{pmatrix} 1 \\ 1 \\ 1 \\ 1 \\ 1 \\ 1 \end{pmatrix} \tag{32}$$

The measurement noise is presented by the dimension matrix $(4 \times 1)$ (33):

$$v_k = \begin{pmatrix} 0.1 \\ 0.1 \\ 0 \\ 0 \end{pmatrix} \tag{33}$$

So the noise and measurement process covariances are deduced from $w_{k-1}$ and $v_k$ by matrices (34) and (35):

$$Q = \begin{pmatrix} 1 & 0 & 0 & 0 & 0 & 0 \\ 0 & 1 & 0 & 0 & 0 & 0 \\ 0 & 0 & 1 & 0 & 0 & 0 \\ 0 & 0 & 0 & 1 & 0 & 0 \\ 0 & 0 & 0 & 0 & 1 & 0 \\ 0 & 0 & 0 & 0 & 0 & 1 \end{pmatrix} \tag{34}$$

$$R = \begin{pmatrix} 0.1 & 0 & 0 & 0 \\ 0 & 0.1 & 0 & 0 \\ 0 & 0 & 0 & 0 \\ 0 & 0 & 0 & 0 \end{pmatrix} \tag{35}$$

### 3.2 The update equations

Finally, the output equations for the two prediction and correction blocks of the Kalman filter are:

- Prediction Eqs. (36) and (37):

$$\hat{s}_k = A \times \hat{s}_{k-1} \tag{36}$$

$$P_k^- = A \times P_{k-1} \times A^T + Q \tag{37}$$

- Correction Eqs. (38), (39) and (40):

$$K_k = P_k^- \times H^T \times \left(H \times P_k^- \times H^T + R\right)^{-1} \tag{38}$$

$$\hat{s}_k = \hat{s}_k^- + K_k \times \left(z_k - H \times \hat{s}_k^-\right) \tag{39}$$

$$P_k = P_k^- - K_k \times H \times P_k^- \tag{40}$$

With $K_k$ presents the gain of the Kalman filter at the moment $k$, $\hat{s}_k$ the state estimated and predicted at the moment $k$ and $P_k$ is the prediction covariance matrix at time $k$. These three Eqs. (38), (39) and (40) present the output parameters of the Kalman filter. To verify the performance and results of the Kalman filter for estimation the parameters of the object tracking system. We compared the state vector values $(x_c, y_c, W, L, x, y)$ for the sequence video "Foreman," computed for this filter, with the state vector values obtained by the tracking algorithm. These values are grouped in **Table 1**. It can be seen that the Kalman filter has good values of the state vectors, very close to those obtained by tracking algorithm. This proves the efficiency of the Kalman filter in the estimation of state vectors for tracking objects, the same for measurement vectors.

| NFr | x | y | W | L | $x_c$ | $y_c$ | $x_{KF}$ | $y_{KF}$ | $W_{KF}$ | $L_{KF}$ | $x_{cKF}$ | $y_{cKF}$ |
|---|---|---|---|---|---|---|---|---|---|---|---|---|
| 2 | 52 | 63 | 58 | 71 | 87.35 | 91.50 | 69 | 53 | 53 | 63 | 87.19 | 91.49 |
| 10 | 55 | 67 | 58 | 71 | 90.62 | 96.41 | 69 | 55 | 55 | 67 | 89.80 | 95.68 |
| 15 | 55 | 68 | 58 | 71 | 90.05 | 96.93 | 55 | 69 | 55 | 68 | 89.58 | 96.81 |
| 25 | 54 | 65 | 58 | 71 | 89.52 | 94.20 | 56 | 65 | 69 | 56 | 90.04 | 93.27 |
| 35 | 54 | 55 | 58 | 71 | 89.18 | 83.92 | 55 | 55 | 69 | 55 | 89.28 | 83.72 |
| 45 | 48 | 56 | 58 | 71 | 83.45 | 85.26 | 49 | 57 | 69 | 49 | 83.36 | 85.37 |
| 55 | 46 | 63 | 58 | 71 | 81.56 | 92.43 | 46 | 64 | 69 | 46 | 80.97 | 92.44 |
| 65 | 43 | 65 | 58 | 71 | 78.52 | 93.53 | 44 | 69 | 69 | 44 | 78.49 | 93.37 |
| 75 | 43 | 63 | 58 | 71 | 78.73 | 92.46 | 43 | 64 | 69 | 43 | 77.98 | 92.59 |
| 85 | 36 | 57 | 58 | 71 | 71.41 | 85.59 | 37 | 57 | 69 | 37 | 71.03 | 85.34 |
| 99 | 38 | 62 | 58 | 71 | 73.80 | 91.03 | 39 | 63 | 69 | 39 | 73.91 | 91.26 |

**Table 1.**
*The results of the state vectors computed by the tracking algorithm (block-matching and Camshift) and the Kalman filter.*

| Video | Nbre | Histogram | Subtraction | Skin color |
|---|---|---|---|---|
| **Sequence** | **Frames** | **Calculation** | **Background** | **Detection** |
| | | **Execution time** | **Execution time** | **Execution time** |
| Foreman | 68 | 16.51 (s) | × | 5.21 (s) |
| Redcup | 68 | 19.21 (s) | × | × |
| Afef | 68 | 17.02 (s) | × | 10.58(s) |
| PETS2001(1) | 68 | 43.54 (s) | 36.56 (s) | × |
| PETS2001(2) | 68 | 47.83 (s) | 43.29 (s) | × |

**Table 2.**
*Precision values calculated for skin detection method and execution times.*

The expression for the estimation error is presented by Eq. (41). We applied this Eq. (41) for all the parameters of the state vectors (x, y, W, L, $x_c, y_c$) of our implemented algorithm for tracking object in a video sequence (Camshift algorithm and Kalman filter algorithm):

$$E_x = x_{Kalman} - x_{Camshift} \tag{41}$$

The results of estimation calculation of the tracking parameters with our algorithm (Camshift and Kalman filter) that we obtain in the different test sequence are given in **Table 2**.

We can see the two pre-treatment methods (background subtraction and skin color detection) are the fastest and the histogram calculation method is the slowest.

## 4. The results for tracking and detection objects

The fundamental basis for estimating the parameters of tracking by the Kalman filter consists in estimating the state vector $s_k$ and the measurement vector $z_k$. These vectors are calculated by the Camshift algorithm. The parameters of the vectors are the center of

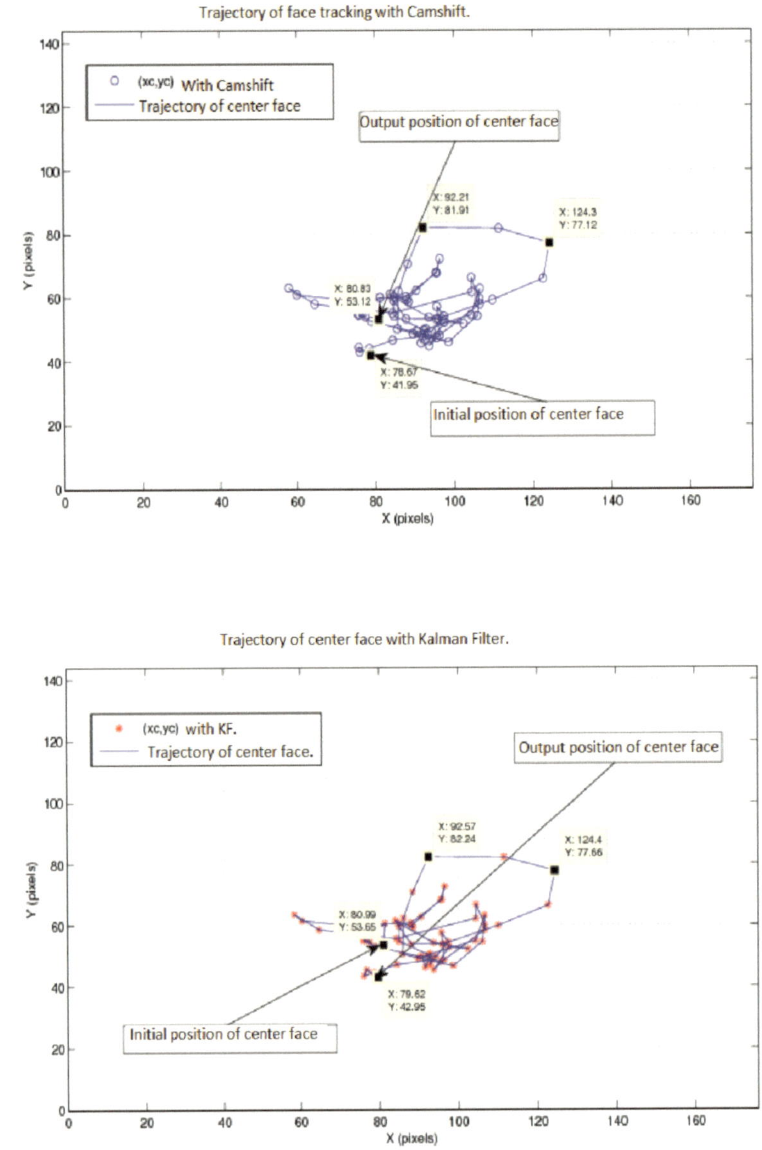


**Figure 5.**
*The trajectory of the gravity center of the face corrected by the Kalman filter.*

gravity of the object $(x_c, y_c)$, the position in each image $(x, y)$ and the width and length of the search window $(L, W)$. **Figure 4** shows the trajectory of the gravity center of the face estimated by the Kalman filter. We can see other results in our publications [1–3], there are presents the prediction and correction of trajectory of an different objects (human, car, glass, mono and multi-objects) in different environments (**Figure 5**).

We can see another correction of trajectory of car (in too sequences video) in **Figure 6**.

## 5. Results discussion

During the test sequences generated with the different pre-processing methods, we can conclude that object tracking differs from one object to another (a human

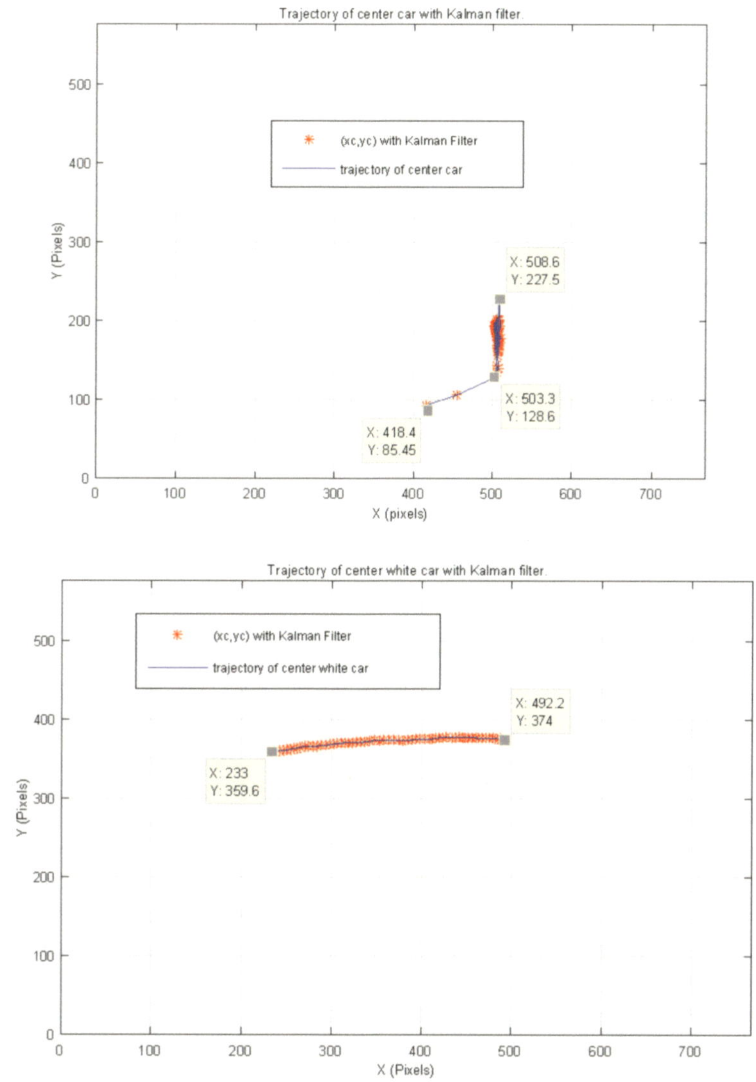


**Figure 6.**
*The trajectory of the gravity center of the face corrected by the Kalman filter.*

being, a face, a hand, a glass, a car) and that several parameters can influence the monitoring result.

The experimental results obtained indicate that our algorithm (Camshift and the Kalman filter) gives superior results, in terms of precision, reliability and execution time, in comparison with the various methods presented in the literature (for example the KLT (Kanade Lucas Tomasi) algorithm and the classifier algorithm (Adaboost and SVM) [1–3, 7]). In particular, the use of several preprocessing methods to detect the object in each frame of the sequence. The results of the implemented algorithms are the meanshift displacement algorithm and block-matching, the Camshift algorithm and the Kalman filter, this combination for the this algorithms give a robust, precise, reliable and fast algorithm.

Evaluating the performance of a mobile object tracking system in a video sequence is a complex task which requires the definition of metrics bringing into play concepts specific to video analysis, such as time persistence, precision and execution time for example.

## 6. Conclusions

The detection and tracking of objects in a sequence of images or video is a topical need for several applications such as video conferencing, video indexing and especially video surveillance. Computer vision with a Human Interface Machine "HIM" is therefore an issue actively studied in many domains, especially since the prices of acquisition and processing equipment have become more attractive. This is an area that touches on everything, starting from the problems of acquisition with different linked effects and where the originality of simple ideas can still bring a lot. In this chapter, we introduced the Kalman filter algorithm for tracking and detection objects and multi-objects. Localization, target tracking, and detection objects were provided as examples for reader's better understanding of practical usage of the Kalman filters. We proceeded to the implementation of the different modules of object tracking algorithm through the estimation of calculation parameters using a Kalman filter. The results obtained make it possible to meet the monitoring requirements of several video surveillance applications. On the one hand, the localization precision achieved by our system makes it a standard module for detection or identification or object tracking systems. On the other hand, a flow at a frequency of 20 frames per second was considered, which is reasonable for an object tracking system with a minimum execution time. The tracking algorithm with its different modules must be tested with other video sequences. Although the implementation of monitoring systems has certain weaknesses, our method has given promising results. Many avenues can be envisaged to continue this work. First of all, note that we tested the algorithm implemented for tracking two objects (a car and a pedestrian in the sequence of "PETS 2001 (1)" and two cars in the sequence of "PETS 2001 (2)"), and it can be applied for tracking multiple objects in a video sequence. Then, use the detection algorithm based on Adaboost classifiers upstream of the tracking algorithm (Camshift and Kalman filter). The association of these two modules is based on a cascade of Adaboost classifiers, improves the calculation time and improves the quality of tracking of one or more objects in a sequence of images or video. Then, validation of the detection and tracking system for faces and other objects (pedestrians, cars, hand gestures, glass, etc.) on an FPGA target platform (Saber-Lite with ARM-Cortex-A9MP). Our solution optimize the time of execution and other criteria in frame and video processing. In future, we intend to extensively evaluate the method quantitatively so that it can be well tested before trying on computer vision practice.

## Author details

Afef Salhi[1*], Fahmi Ghozzi[1,2] and Ahmed Fakhfakh[1,2]

1 Digital Research Center of SFAX (CRNS), Laboratory of Technology for Smart Systems (LT2S), Sfax, Tunisia

2 ENET'COM, University of Sfax, Sfax, Tunisia

*Address all correspondence to: salhiafefge@gmail.com; afef.salhi.ge@enis.tn

## References

[1] Salhi A, Moresly Y, Ghozzi F, Yengui A, Fakhfakh A. Modeling from an object and multi-object tracking system. In: 2016 Global Summit on Computer Information Technology (GSCIT); 16–18 July 2016. Vol. 1. Sousse-Tunisia; 2016. pp. 80-85

[2] Salhi A, Moresly Y, Ghozzi F, Fakhfakh A. Face detection and tracking system with block-matching, Meanshift and Camshift algorithms and Kalman filter. In: 18th International Conference on Sciences and Techniques of Automatic Control Computer Engineering; 21–23 December 2017. Monastir-Tunisia: IEEE; 2017. pp. 139-145

[3] Salhi A, Ghozzi F, Fakhfakh A. Toward a methodology for object tracking system in computer vision. In: 14 th Tunisia-Japan Symposium on Science, Society Technology TJASSST'17; 24–26 November 2017. Gammarth-Tunisia; 2017. pp. 185-187

[4] Nor Nadirah AA, Mostafah YM, Shafie AW, Zainuddin NA, Rashidan MA. Features-based moving objects tracking for smart video surveillances: A review. International Journal on Artificial Intelligence Tools. 2018;**27**(2):1830001

[5] Chavda HK, Dhamecha M. Moving object tracking using PTZ camera in video surveillance system. In: International Conference on Energy, Communication, Data Analytics and Soft Computing (ICECDS-2017). IEEE; 2017. pp. 263-266

[6] Mueller TM, Karasev P, Kolesov I, Tannenbaum A. Optical flow estimation for flame detection in videos. In: IEEE Transactions on Image Processing. Vol. 22. 2013. pp. 1-6

[7] Salhi A. Study and implementation a system tracking and detection objects in video sequence. In: 2011 at National Engineering School of Sfax "ENIS". Sfax-Tunisia; 21 July 2011. pp. 1-100

[8] Deepak P, Krishnakumar S, Suresh S. Human recognition for surveillance systems using bounding box. In: International Conference on Contemporary Computing and Informatics Conference (IC3I). IEEE; 2014. pp. 852-856

[9] Sakai Y, Oda T, Ikeda M, Barolli L. An object tracking system based on SIFT and SURF feature extraction methods. In: 18th International Conference on Network-Based Information Systems (ICN-BIS), 978-1-4799-9942-2/15-CPS. IEEE; 2015. pp. 561-565. DOI: 10.1109/NBiS.2015.121

[10] Du D, Qi Y, Yu H, Yang Y, Duan K, Lu G, et al. The Unmanned Aerial Vehicle Benchmark: Object Detection and Tracking. Switzerland: Springer Nature; 2018. p. 375391. DOI: 10.1007/978-3-030-01249-623

Chapter 5

# Kalman Filtering Applied to Induction Motor State Estimation

*Yassine Zahraoui and Mohamed Akherraz*


## Abstract

This chapter presents a full definition and explanation of Kalman filtering theory, precisely the filter stochastic algorithm. After the definition, a concrete example of application is explained. The simulated example concerns an extended Kalman filter applied to machine state and speed estimation. A full observation of an induction motor state variables and mechanical speed will be presented and discussed in details. A comparison between extended Kalman filtering and adaptive Luenberger state observation will be highlighted and discussed in detail with many figures. In conclusion, the chapter is ended by listing the Kalman filtering main advantages and recent advances in the scientific literature.


**Keywords:** Kalman filtering, stochastic algorithm, non-linear discrete system, state variables estimation, standard Kalman filter, extended Kalman filter

## 1. Introduction

Kalman filtering is an algorithm that employs a series of observations over time, containing noise and other inaccuracies, and generates approximations of unknown variables that tend to be more accurate than those based on a single measurement alone, by estimating a joint probability distribution over the variables for each timeframe, as in [1].

The Kalman filter is a state observer which detects the presence of measurement noises as well as uncertainties about an unknown dynamic state system, this system is generally assimilated to state noise by stochastic algorithms tending to minimise the variance of the estimation error, as described in [2].

The Kalman filter is suitable for recursive linear filtering of discrete data. It provides an estimation of a state vector or a parameter and its error covariance and variance matrix that contain information about the accuracy of its state variables, as in [3]. The natural presence of noise when an induction machine is driven by an inverter represents a strong argument for the choice of this kind of observers. Its characteristics will relate to the observation of the speed and the components of the rotor fluxes. The only needed measurements are the stator currents. Some state variables will be provided directly by the control law. Thus, the stator voltages will be considered as inputs for the filter. **Table 1** shows a technical comparison between the adaptive observer and the stochastic filter.

| Technical comparison | | |
|---|---|---|
| **Features** | **Adaptive Luenberger Observer** | **Extended Kalman Filter** |
| Response time | Very well | Very well |
| Tracking error | Little | Very little |
| Torque ripples | Very high | Medium |
| Robustness | Robust | Very robust |
| Noise sensitivity | Very sensitive | Not sensitive |

[a]*The comparison is based on the obtained results.*

**Table 1.**
*Comparison between ALO and EKF.*

## 2. Induction motor parts, features and mathematical model

### 2.1 Induction motor parts and features

Induction motors are the most commonly used electrical machines, they are cheaper, rugged and easier to maintain compared to other alternatives. It has two main parts: stator and rotor, stator is a stationary part and rotor is the rotating part. Stator is made by stacking thin slotted highly permeable steel lamination inside a steel or cast iron frame, winding passes through slots of stator. When a three phase AC current passes through it, something very interesting happens. It produces a rotating magnetic field, the speed of rotation of a magnetic field is known as synchronous speed.

It is called an induction motor because electricity is inducted in rotor by magnetic induction rather than direct electric connection. To collapse such electric magnetic induction, to aid such electromagnetic induction, insulated iron core lamina are packed inside the rotor, such small slices of iron make sure that Eddy current losses are minimal. And this is another big advantage of three phase induction motors.

The parts of a squirrel cage induction motor are shown in **Figure 1**.

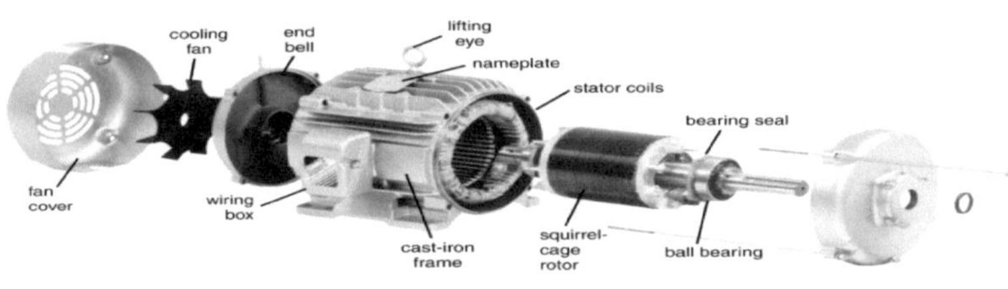


**Figure 1.**
*Squirrel cage induction motor parts.*

### 2.2 Induction motor mathematical model

The induction motor has many state space mathematical models; each model is expressed by assuming a certain state vector. The modelling of AC machines is based mainly on the work of G. Kron, who gave birth to the concept of generalised machine as described in [4]. Park's model is a special case of this concept. It is often used for the synthesis of control laws and estimators. Described by a non-linear algebra-differential system, Park's model reflects the dynamic behaviour of the electrical and electromagnetic modes of the asynchronous machine. It admits several classes of state representations. These model classes depend directly on the control objectives (torque, speed, position), the nature of the power source of the work repository and the choice of state vector components (flux or currents, stator or rotor).

In this chapter, the mathematical model of the machine in use is described in the stator fixed reference frame $(\alpha, \beta)$ (stationary frame) by assuming the stator currents and the rotor fluxes as state variables:

$$\begin{cases} \dot{X} = \mathbf{A}.X + \mathbf{B}.U \\ Y = \mathbf{C}.X \end{cases} \tag{1}$$

Where $X$, $U$ and $Y$ are the state vector, the input vector and the output vector, respectively:

$$X = \left[i_{s\alpha}\ i_{s\beta}\ \phi_{r\alpha}\ \phi_{r\beta}\right]^t;\ U = \left[u_{s\alpha}\ u_{s\beta}\right]^t;\ Y = \left[i_{s\alpha}\ i_{s\beta}\right]^t \tag{2}$$

$$\mathbf{A} = \begin{bmatrix} -\lambda & 0 & \frac{K}{T_r} & K\omega_r \\ 0 & -\lambda & -K\omega_r & \frac{K}{T_r} \\ \frac{L_m}{T_r} & 0 & -\frac{1}{T_r} & -\omega_r \\ 0 & \frac{L_m}{T_r} & \omega_r & -\frac{1}{T_r} \end{bmatrix};\ \mathbf{B} = \begin{bmatrix} \frac{1}{\sigma L_s} & 0 \\ 0 & \frac{1}{\sigma L_s} \\ 0 & 0 \\ 0 & 0 \end{bmatrix};\ \mathbf{C} = \begin{bmatrix} 1 & 0 & 0 & 0 \\ 0 & 1 & 0 & 0 \end{bmatrix} \tag{3}$$

With:

$$\lambda = \frac{R_s}{\sigma.L_s} + \frac{1-\sigma}{\sigma.T_r}; K = \frac{1-\sigma}{\sigma.L_m}; \sigma = 1 - \frac{L_m^2}{L_s.L_r}; T_r = \frac{L_r}{R_r}. \tag{4}$$

The rotor motion is expressed by:

$$J.\frac{d\Omega_r}{dt} = T_{em} - T_L - f.\Omega_r \tag{5}$$

Where $J$ is the motor inertia, $T_{em}$ is the electromagnetic torque, $T_L$ is the load torque, and $f$ is the friction coefficient.

**Figure 2** shows the state space mathematical model of an induction motor.

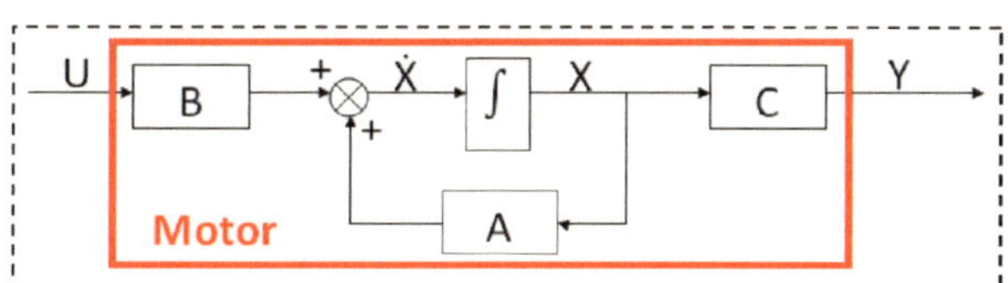


**Figure 2.**
*Induction motor state space mathematical model.*

## 3. Standard Kalman filter

In this chapter, the process to be observed is an induction motor. Its state is composed of stator currents and rotor fluxes in $\alpha$-$\beta$ reference frame, the motor model and its components are shown in **Figure 3**. The motor model is defined by a

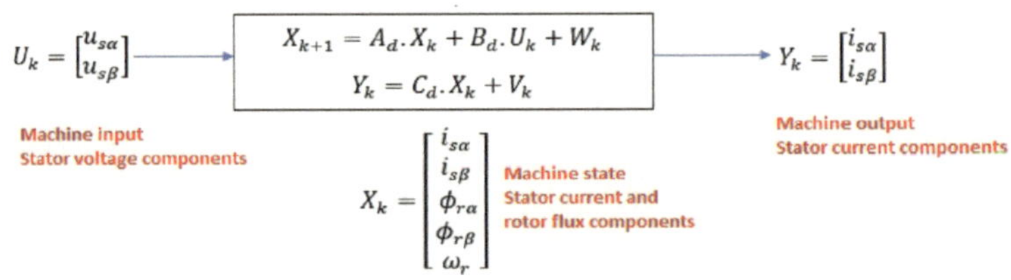


**Figure 3.**
*Induction motor model with input, state and output components.*

discrete time linear state model composed of two additional terms for taking into account discrete state noise $W_k$ and discrete state measurement $V_k$.

$$\begin{cases} X_{k+1} = A_d \cdot X_k + B_d \cdot U_k + W_k \\ Y_k = C_d \cdot X_k + V_k \end{cases} \tag{6}$$

The addition of noise is necessary, since the noise-free equations (deterministic model) define an ideal system. A more realistic model (stochastic model) is obtained by adding the noise vectors. Some assumptions are made about discrete noises: they are white, Gaussian, their average is zero and they are correlated neither with each other nor with the state variables. These properties derive the following equations:

$$E\{W_k\} = 0; E\{V_k\} = 0; E\{W \cdot W^t\} = Q; E\{V \cdot V^t\} = R\cdot \tag{7}$$

$$E\{W_k \cdot V_{k-\tau}^t\} = 0; E\{W_k \cdot X_{k-\tau}^t\} = 0; E\{V_k \cdot X_{k-\tau}^t\} = 0\cdot \tag{8}$$

$E$ represents the expectation value operator.

The implementation of the Kalman filter algorithm requires two phases, the first one is a prediction phase which consists in determining the prediction vector $X_{k+1|k}$ from the process state equations and also the previous values of the estimated states $X_{k|k}$ at time $k$. In addition, the predicted state covariance matrix $P$ is also obtained before the new measurements are made, for this purpose the mathematical model and also the covariance matrix $Q$ of the system are used.

$$X_{k+1|k} = A_d \cdot X_{k|k} + B_d \cdot U_k \tag{9}$$

$$P_{k+1|k} = A_d \cdot P_{k|k} \cdot A_d^t + Q \tag{10}$$

The second phase then consists of the correction. It consists in correcting the prediction vector by the measurement vector by adding a correction term $K \cdot$

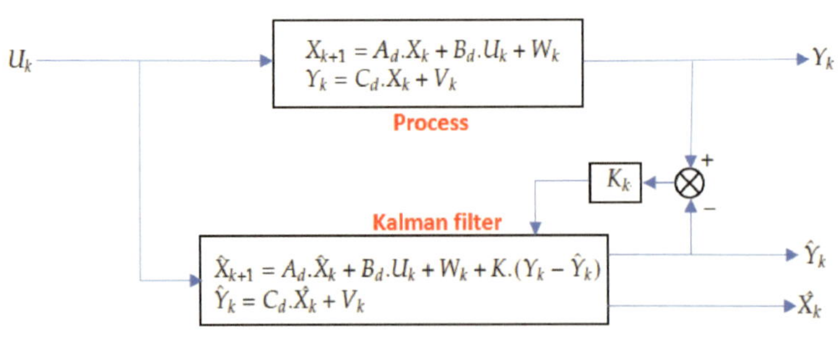


**Figure 4.**
*Standard Kalman filter principle.*

$(Y - \hat{Y})$ to the predicted states $X_{k+1|k}$ obtained in the first phase. This correction term is a weighted difference between the actual output vector $Y$ and the predicted output vector $\hat{Y}$. Thus, the predicted state estimate and also its covariance matrix are corrected by a feedback correction system to obtain the estimate of the state vector $X_{k+1|k}$ at the present moment $k + 1$. **Figure 4** below shows the principle of the standard Kalman filter.

$$K_{k+1} = P_{k+1|k} \cdot C_d^t \cdot \left[C_d \cdot P_{k+1|k} \cdot C_d^t + R\right]^{-1} \tag{11}$$

$$X_{k+1|k+1} = X_{k+1|k} + K_{k+1} \cdot \left[Y_{k+1} - C_d \cdot X_{k+1|k}\right] \tag{12}$$

$$P_{k+1|k+1} = P_{k+1|k} - K_{k+1} \cdot C_d \cdot P_{k+1|k} \tag{13}$$

Where $K$ denotes the gain matrix of the Kalman filter, $P$ is the estimation error covariance matrix, $Q$ and $R$ are, respectively, the covariance matrices of the state and the measurement noises. The gain matrix $K$ is chosen so as to minimise the variance of the estimation error. This minimization will focus on the diagonal elements of the estimation matrix. Thus, the Kalman filter algorithm uses on one hand the knowledge of the process to predict the state vector, and on other hand the actual measurements to correct the predicted vector. The standard Kalman filter previously described allows estimation of the state of a linear system. If we want to estimate an additional parameter outside the state vector, as the rotational speed of an induction motor, one solution is to extend the estimated state vector to the speed of rotation. The model then becomes non-linear and in this case, the extended Kalman filter is required.

## 4. The extended Kalman filter

The extended Kalman filter performs an estimation of the state of a non-linear process. It allows in particular to add, to the state vector, another variable that we wish to estimate. This filter is widely used for estimating the various quantities of the induction machine, such as: rotor speed, load torque, electrical and mechanical parameters. Given that the extended Kalman filter is only the application of the standard Kalman filter previously described in the case of a non-linear system, it is then necessary to perform a linearization of this system at each step around the operating point defined in the previous step. Let the non-linear model of the system to be observed defined by the following equation of state:

$$\begin{cases} X_{e_{k+1}} = f\left(X_{e_k}, U_k\right) + W_k \\ Y_k = h\left(X_{e_k}\right) + V_k \end{cases} \tag{14}$$

Such as:

$$X_{e_k} = [X_k \;\; \Theta_k]^t. \tag{15}$$

Where $f$ and $h$ are non-linear functions, $X_{e_k}$ designates the extended state vector, $X_k$ and $\Theta_k$ are considered, respectively, as the main state vector and the parameter vector (composed of parameters and unknown inputs to be estimated). These parameters vary very little with respect to other quantities, for that reason we put $\Theta_{k+1} = \Theta_k$. The following discrete augmented state model is constructed:

$$\begin{bmatrix} X_{k+1} \\ \Theta_{k+1} \end{bmatrix} = \begin{bmatrix} A_d & 0 \\ 0 & I \end{bmatrix} \begin{bmatrix} X_k \\ \Theta_k \end{bmatrix} + \begin{bmatrix} B_d \\ 0 \end{bmatrix} U_k + W_k \tag{16}$$

$$Y_k = \begin{bmatrix} C_d & 0 \end{bmatrix} \begin{bmatrix} X_k \\ \Theta_k \end{bmatrix} + V_k \tag{17}$$

$A_d$, $B_d$ and $C_d$ are, respectively, the state, the input and the discrete output matrices. $I$ is the identity matrix.

The implementation of the extended Kalman filter algorithm to the discrete non-linear system requires the execution of the following steps:

- Initialization of the states $X_{0|0}$, the parameters $\Theta_{0|0}$ and the covariance matrices $P_{0|0}$, $Q$ and $R$

- Prediction of the states and the parameters

$$X_{k+1|k} = A_d \cdot X_{k|k} + B_d \cdot U_k \tag{18}$$

$$\Theta_{k+1|k} = \Theta_{k|k} \tag{19}$$

- Prediction of error covariance matrix

$$P_{k+1|k} = F_k \cdot P_{k|k} \cdot F_k^t + Q \tag{20}$$

$F_k$ is the gradient matrix defined as follows:

$$F_k = \left.\frac{\partial f\left(X_{e_k}, U_k\right)}{\partial X_{e_k}}\right|_{X_{e_{k|k}}} = \begin{bmatrix} A_d & \left.\frac{\partial}{\partial \Theta_k}\left(A_d \cdot X_{k|k} + B_d \cdot U_k\right)\right|_{\Theta_{k|k}} \\ 0 & I \end{bmatrix} \tag{21}$$

- Calculation of Kalman gain

$$K_{k+1} = P_{k+1|k} \cdot H_k^t + \left[H_k \cdot P_{k+1|k} \cdot H_k^t + R\right]^{-1} \tag{22}$$

$H_k$ is the gradient matrix defined as follows:

$$H_k = \left.\frac{\partial h\left(X_{e_k}\right)}{\partial X_{e_k}}\right|_{X_{e_{k|k}}} = \begin{bmatrix} C_d & \left.\frac{\partial}{\partial \Theta_k}\left(C_d \cdot X_{k|k}\right)\right|_{\Theta_{k|k}} \\ 0 & I \end{bmatrix} \tag{23}$$

- Estimation of the states and the parameters

$$\begin{bmatrix} X_{k+1|k+1} \\ \Theta_{k+1|k+1} \end{bmatrix} = \begin{bmatrix} X_{k+1|k} \\ \Theta_{k+1|k} \end{bmatrix} + K_{k+1} \cdot \left[Y_{k+1} - C_d \cdot X_{k+1|k}\right] \tag{24}$$

- Estimate of the error covariance matrix

$$P_{k+1|k+1} = P_{k+1|k} - K_{k+1} \cdot K_k \cdot P_{k+1|k} \tag{25}$$

- Update matrices at instant $k = k + 1$

$$X_{k|k} = X_{k+1|k+1}$$
$$\Theta_{k|k} = \Theta_{k+1|k+1} \tag{26}$$
$$P_{k|k} = P_{k+1|k+1}$$

# 5. Application to the estimation of induction machine speed and flux

## 5.1 Induction machine extended model

The continuous model of the induction machine extended to the electrical rotational speed is represented by a non-linear system of state equations:

$$\begin{cases} \dot{X}_e(t) = f(X_e(t), U(t)) = A \cdot X_e(t) + B \cdot U(t) \\ Y(t) = h(X_e(t)) = C \cdot X_e(t) \end{cases} \tag{27}$$

In which:

$$X_e = [X \quad \Theta]^t = \begin{bmatrix} i_{s\alpha} & i_{s\beta} & \phi_{r\alpha} & \phi_{r\beta} & \omega_r \end{bmatrix}^t \; Y = \begin{bmatrix} i_{s\alpha} & i_{s\beta} \end{bmatrix}^t U = \begin{bmatrix} u_{s\alpha} & u_{s\beta} \end{bmatrix}^t \tag{28}$$

With:

$$A = \begin{bmatrix} -\gamma & 0 & \frac{\mu}{T_r} & \mu \cdot \omega_r & 0 \\ 0 & -\gamma & -\mu \cdot \omega_r & \frac{\mu}{T_r} & 0 \\ \frac{L_m}{T_r} & 0 & 1 - \frac{1}{T_r} & -\omega_r & 0 \\ 0 & \frac{L_m}{T_r} & \omega_r & \frac{1}{T_r} & 0 \\ 0 & 0 & 0 & 0 & 0 \end{bmatrix} B = \begin{bmatrix} \frac{1}{\sigma \cdot L_s} & 0 \\ 0 & \frac{1}{\sigma \cdot L_s} \\ 0 & 0 \\ 0 & 0 \\ 0 & 0 \end{bmatrix} \tag{29}$$

## 5.2 Discretization of the continuous model

The previous model of the induction machine must be discretized for the implementation of the extended Kalman filter. If quasi-constant control voltages are assumed over a sampling period $T_s$ as in [5], the discrete augmented state model can be approximated by:

$$\begin{cases} X_{e_{k+1}} = f\left(X_{e_k}, U_k\right) = A_d \cdot X_{e_k} + B_d \cdot U_k \\ Y_k = h\left(X_{e_k}\right) = C_d \cdot X_{e_k} \end{cases} \tag{30}$$

The matrices of this model are obtained by a limited development in Taylor series of order one:

$$A_d \approx e^{A \cdot T_s} = I + A \cdot T_s; B_d = B \cdot T_s; C_d = C \cdot \tag{31}$$

This leads to:

$$A_d = \begin{bmatrix} 1 - T_s \cdot \gamma & 0 & T_s \cdot \frac{\mu}{T_r} & T_s \cdot \mu \cdot \omega_r & 0 \\ 0 & 1 - T_s \cdot \gamma & -T_s \cdot \mu \cdot \omega_r & T_s \cdot \frac{\mu}{T_r} & 0 \\ T_s \cdot \frac{L_m}{T_r} & 0 & 1 - \frac{T_s}{T_r} & -T_s \cdot \omega_r & 0 \\ 0 & T_s \cdot \frac{L_m}{T_r} & T_s \cdot \omega_r & 1 - \frac{T_s}{T_r} & 0 \\ 0 & 0 & 0 & 0 & 1 \end{bmatrix}; B_d = \begin{bmatrix} T_s \cdot \frac{1}{\sigma \cdot L_s} & 0 \\ 0 & T_s \cdot \frac{1}{\sigma \cdot L_s} \\ 0 & 0 \\ 0 & 0 \\ 0 & 0 \end{bmatrix} \tag{32}$$

### 5.3 Implementation of the extended Kalman filter to the induction machine discrete system

The application of the extended Kalman filter to the discrete system of the induction machine, taking into account the presence of state noise $W_k$ and measurement noise $V_k$. This leads to the following expressions:

$$\begin{cases} X_{e_{k+1}} = f(X_{e_k}, U_k) + W_k = A_d \cdot X_{e_k} + B_d \cdot U_k + W_k \\ Y_k = h(X_{e_k}) + V_k = C_d \cdot X_{e_k} + V_k \end{cases} \tag{33}$$

With:

$$X_{e_{k+1}} = [X_k \quad \Theta_k]^t; \; Y_k = \left[i_{s\alpha_k} \quad i_{s\beta_k}\right]^t; \; U_k = \left[u_{s\alpha_k} \quad u_{s\beta_k}\right]^t; \; W_k = \left[W_{x_k} \quad W_{\Theta_k}\right]^t \tag{34}$$

Similarly, the linearization matrix $H_k$ is written as follows:

$$H_k = \begin{bmatrix} 1 & 0 & 0 & 0 & 0 \\ 0 & 1 & 0 & 0 & 0 \end{bmatrix} \tag{35}$$

In the determination of the initial covariance matrix $P_{0|0}$, it is generally limited to the choice of elements on the diagonal. These elements are chosen in such a way that they correspond to the uncertainty about estimates of initial state variables.

#### *5.3.1 Choice of covariance matrices Q and R*

It is via these matrices that the various measured, predicted and estimated states will pass. Their goals are to minimise the errors associated with approximate modelling and the presence of noise on the measurements. This is the most difficult point of applying the Kalman filter to observation. The matrix $Q$ linked to the noises tainting the state, allows adjusting the estimated quality of the modelling and discretization. A strong value of $Q$ gives a high value of the gain K stimulating the importance of the modelling and the dynamics of the filter. A high value of $Q$ can, however, create an instability of the observation. The matrix $R$ regulates the weight of the measurements. A high value indicates a great uncertainty of the measurement. On the other hand, a low value makes it possible to give a significant weight to the measurement.

#### *5.3.2 The reference recursive recipe (RRR) method for the EKF*

We can consider the choice of the Kalman filter calibration matrices $Q$ and $R$, as well as the initial values of the estimated state vector $X_{e_{0|0}}$ and the matrix $P_{0|0}$, as degrees of freedom of the Kalman filter. The following steps explain the recursive or iterative RRR algorithm for the EKF [6]:

- Given the system model and the measurements, the first filter pass through the data of EKF is carried out using guess values of $X_{e_{0|0}}$, $P_{0|0}$, $\Theta$, $R$ and $Q$.

- The RTS smoother is used backwards to get smoothed state and covariance estimates.

- If $X_{e_{0|0}}$ is unknown, then the smoothed state values can be used as the initial state values.

- The estimated smoothed $P_{0|0}$ is scaled up by the number of time points $N$ and further all elements except the diagonal terms corresponding to the parameters are set to zero. Due to the effect of statistical percolation effect, the estimated $R$ and $Q$ will in general be full. But, only the diagonal terms in $Q$ need to be used in the basic state equations and not in the parameter states. Only the diagonal terms in $R$ need to be used in the measurement equations. These are summarised as below. The quadrant on the upper left denotes the state, the bottom right the parameter states, and the others the cross terms.

$$X_{e_{0|0}} = \begin{bmatrix} 0 & 0 & 0 & 0 & 0 \end{bmatrix}^t; P_{0|0} = \begin{bmatrix} 1 & 0 & 0 & 0 & 0 \\ 0 & 1 & 0 & 0 & 0 \\ 0 & 0 & 1 & 0 & 0 \\ 0 & 0 & 0 & 1 & 0 \\ 0 & 0 & 0 & 0 & 1 \end{bmatrix}; \tag{36}$$

$$Q = \begin{bmatrix} q_{11} & 0 & 0 & 0 & 0 \\ 0 & q_{22} & 0 & 0 & 0 \\ 0 & 0 & q_{33} & 0 & 0 \\ 0 & 0 & 0 & q_{44} & 0 \\ 0 & 0 & 0 & 0 & q_{55} \end{bmatrix}; R = \begin{bmatrix} r_{11} & 0 \\ 0 & r_{22} \end{bmatrix} \tag{37}$$

- Then, the filter is run again using the above updates of $X_{e_{0|0}}$, $P_{0|0}$, $\Theta$, $Q$ and $R$ till statistical equilibrium is reached.

**Figure 5** illustrates the state space mathematical model of the observer.

### *5.3.3 Simulation results*

In this section, an extended Kalman filter is implemented in an induction motor vector control scheme. The EKF is designed to observe the motor states: the *d-q* stator phase current components $i_{ds}$, $i_{qs}$, the *d-q* rotor flux components $\phi_{dr}$, $\phi_{qr}$ and the mechanical speed $\omega_r$. The control law and the observer are implemented in MatLab/Simulink software. A load torque of $+10\ N \cdot m$ is applied at $t_1$=0.6$s$ and removed at $t_2$=1.6$s$ in order to show the system robustness against the external perturbation. **Table A1** lists the parameters of the machine used in simulation.

The torque reference $T^*_{em}$ is generated by the speed controller, while the stator voltage references $V^*_{ds}$ and $V^*_{qs}$ are generated by the stator current controllers.

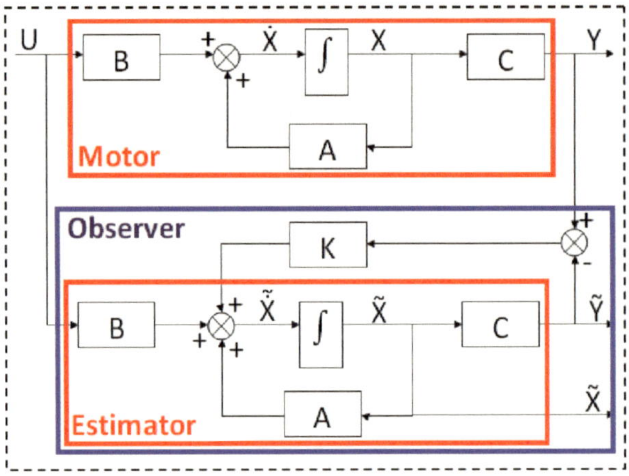


**Figure 5.**
*The observer state space mathematical model.*

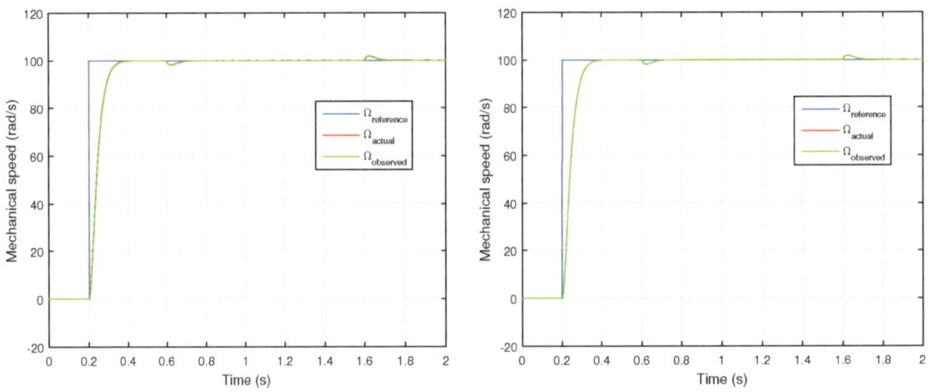


**Figure 6.**
*Speed response.*

The observed speed $\hat{\Omega}_r$ and rotor flux $\hat{\phi}_r$ are served for speed and flux regulators [7, 8]. The coordinate transformation generates the *abc* components needed by the PWM modulator.

The slip frequency is delivered by an integrator, this slip is the most important parameter for the indirect vector schemes. it depends on the observed rotor flux generated by the EKF observer.

**Figures 6–19** illustrate a performance comparison between the two observers: EKF in the left and ALO in the right. **Figure 7** shows the speed response according to the step speed reference of +100 $rad/s$. Both observers show good dynamic at starting up and the speed regulation loop rejects the applied load disturbance quickly. The two observers kept the same fast speed response since the same PI speed controller is used for both speed loops, there is no difference in the transient response. The system response time is very quick and does not exceed $0.2s$, the sufficient time to achieve the permanent regime.

**Figure 8** shows the rotor flux response, it achieves the reference which is 1 $Wb$ very quickly. Even the step speed reference starts at $0.2\,s$, the rotor flux response is independent to the speed application. It must reach the reference very rapidly at the starting up. Then, **Figure 9** shows the torque responses with the load application. At

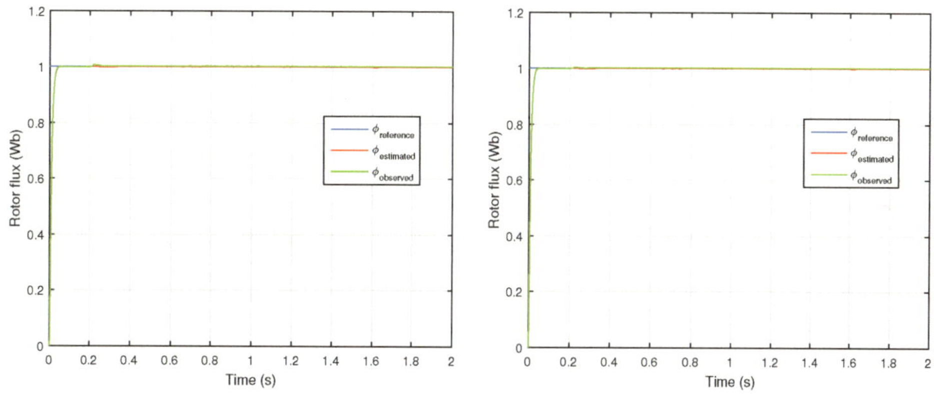


**Figure 7.**
*Rotor flux response.*

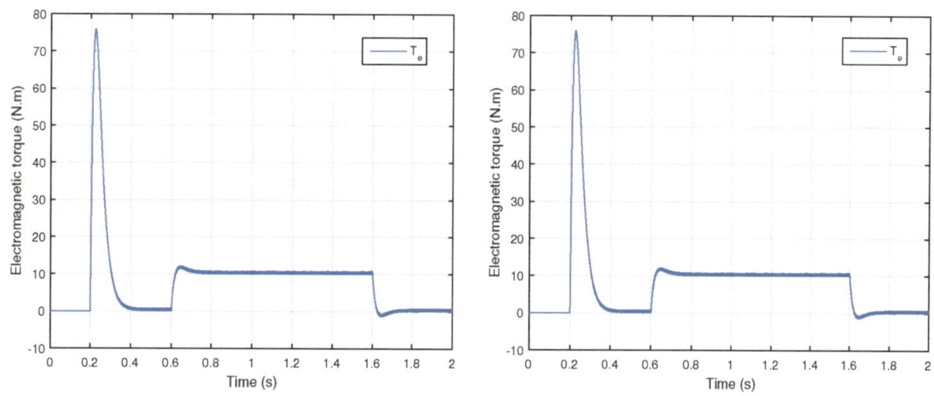


**Figure 8.**
*Electromagnetic torque response.*

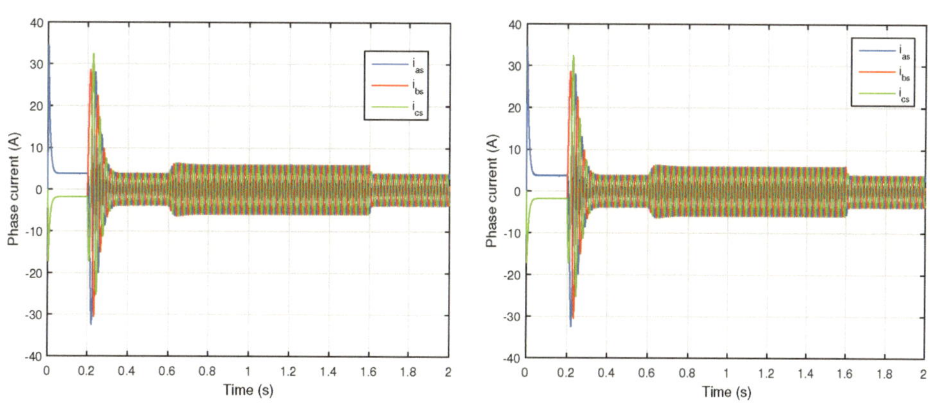


**Figure 9.**
*Stator phase currents response.*

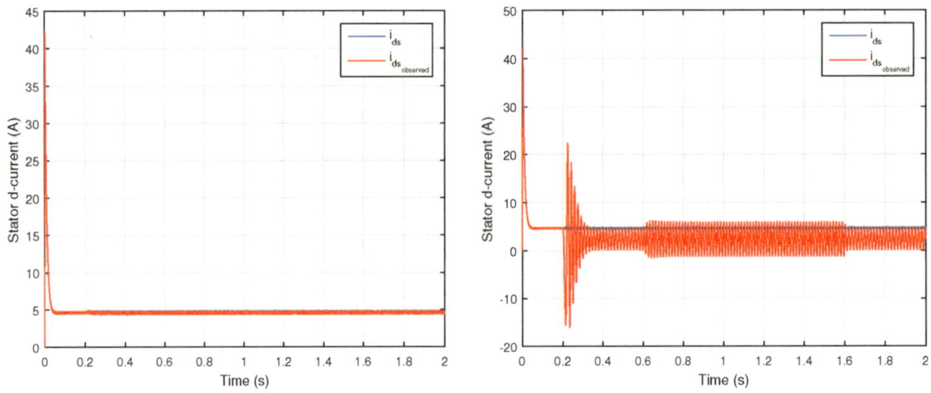


**Figure 10.**
*$i_{ds}$ current response.*

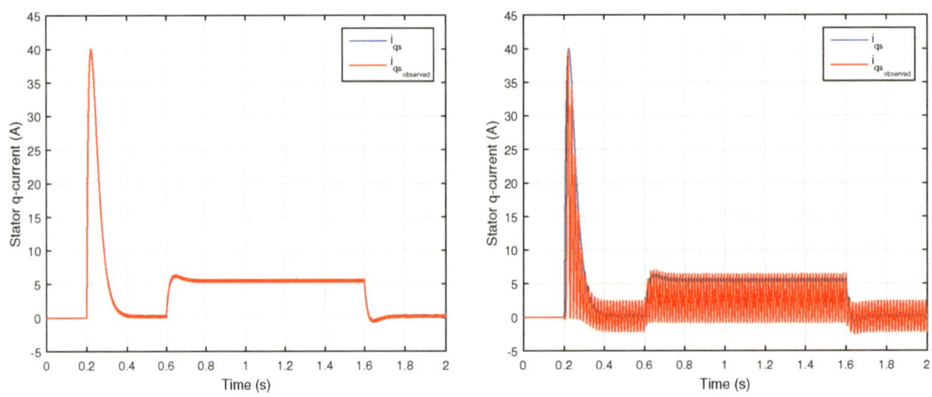


**Figure 11.**
*$i_{qs}$ current response.*

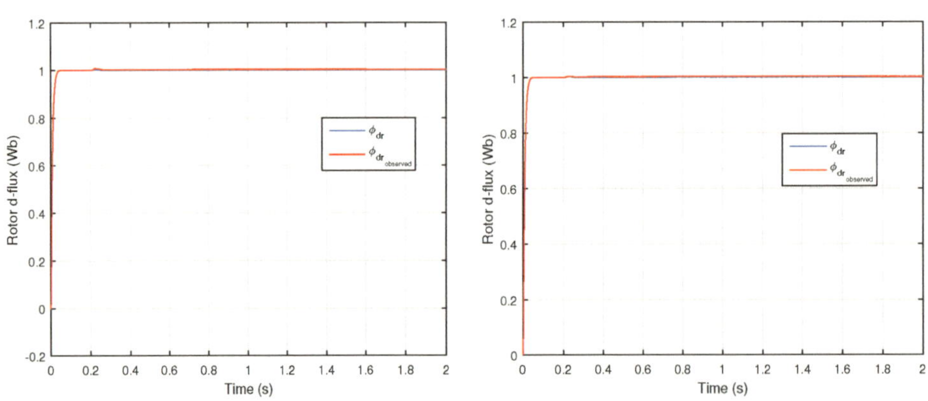


**Figure 12.**
*$\phi_{dr}$ flux response.*

DOI: http://dx.doi.org/10.5772/intechopen.92871

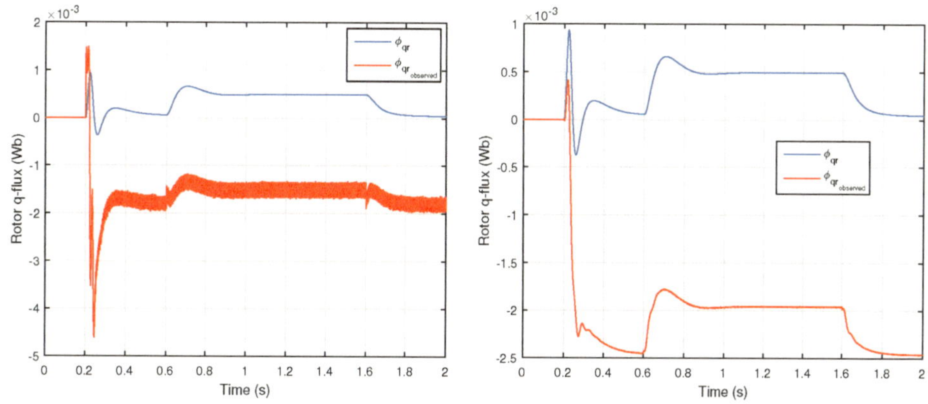


**Figure 13.**
*$\phi_{qr}$ flux response.*

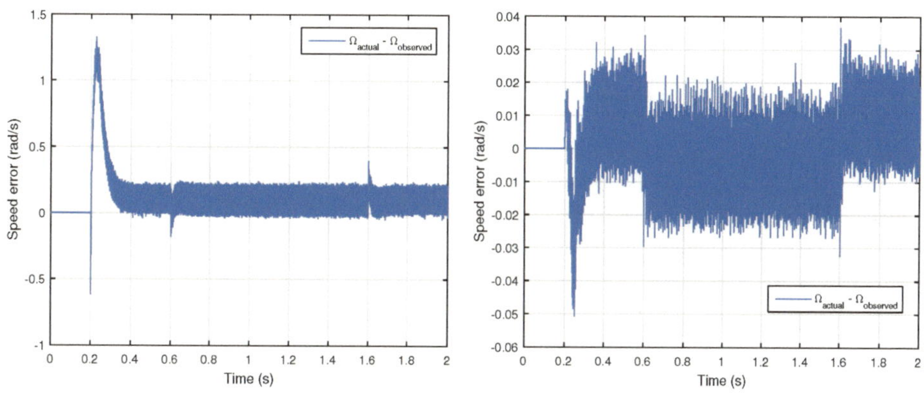


**Figure 14.**
*Mechanical speed error.*

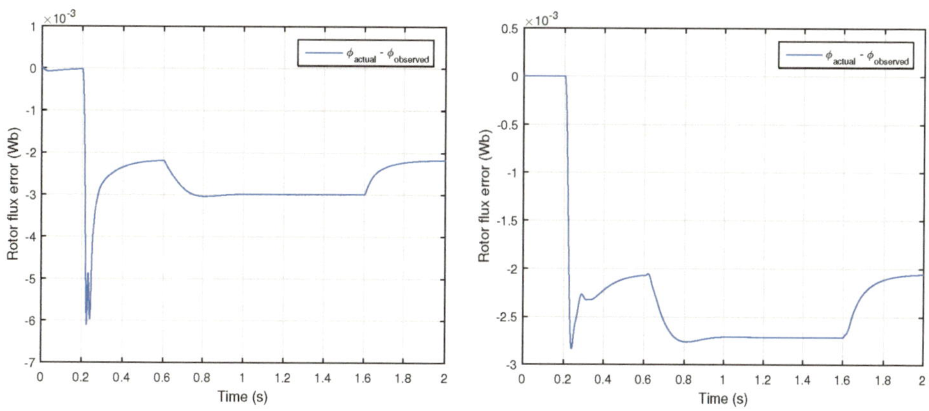


**Figure 15.**
*Rotor flux error.*

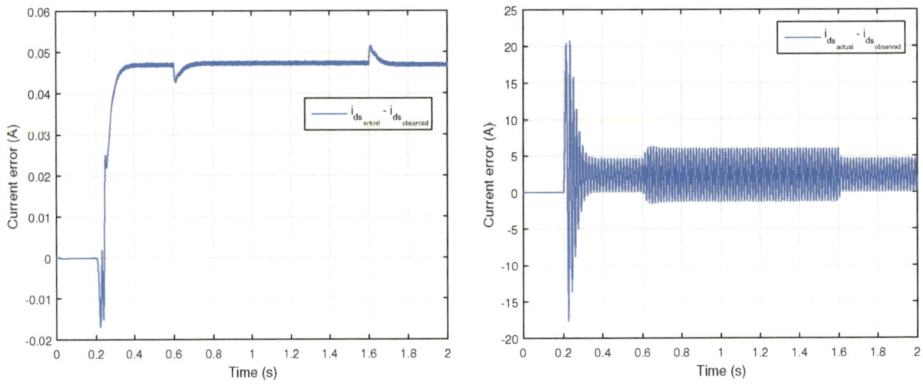


**Figure 16.**
*$i_{ds}$ component error.*

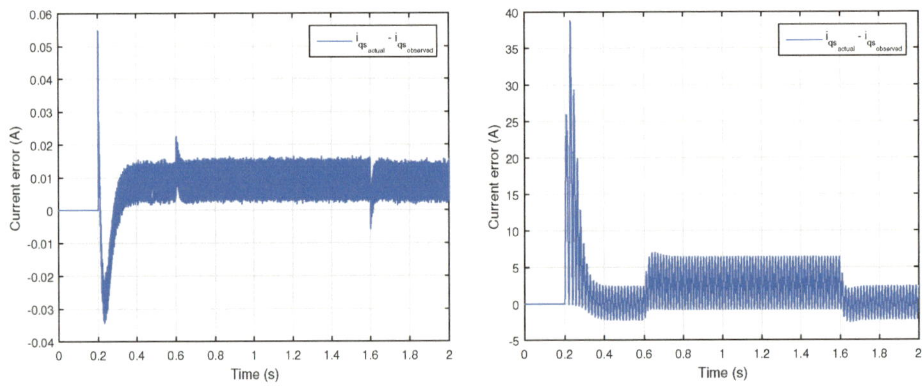


**Figure 17.**
*$i_{qs}$ component error.*

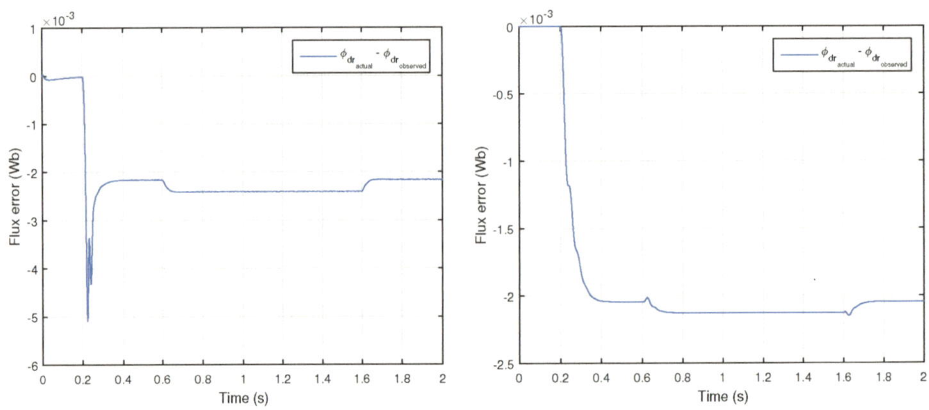


**Figure 18.**
*$\phi_{dr}$ component error.*

the beginning, the speed controller operates the system at the physical limit since the step reference is the hardest for most control processes.

Until now, no apparent difference in the performance of the two observers, **Figures 11–19** will reveal this difference. **Figures 11** and **12** illustrate, respectively, the observed stator current components $i_{ds}$ and $i_{qs}$. We can notice clearly the

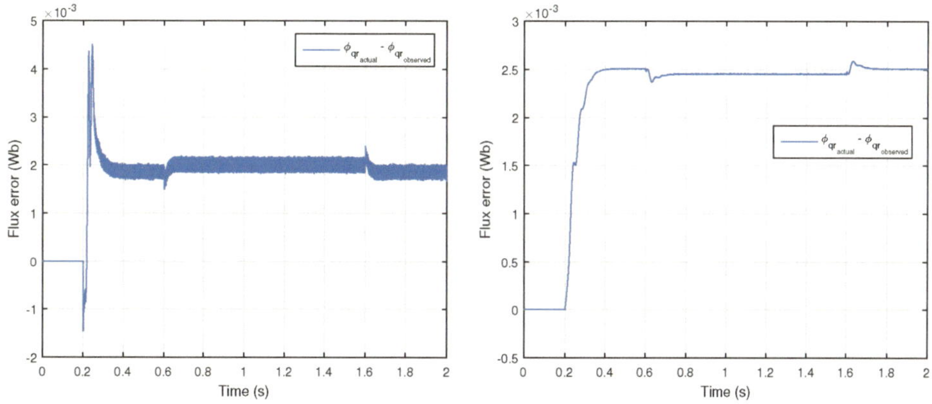


**Figure 19.**
$\phi_{qr}$ *component error.*

superiority of the EKF, no fluctuations seen around the reference. EKF uses a series of measurements containing noise and other inaccuracies contrary to ALO that employs only free noise measurements. **Figures 13** and **14** illustrate, respectively, the observed rotor flux components $\phi_{dr}$ and $\phi_{qr}$. No fluctuations seen around the reference for both observers, only a small static error of observation. Finally, **Figures 15–19** illustrate the static error of all the observed components: the machine state parameters, the rotor flux and the mechanical speed.

All the quantities observed by the EKF are filtered and precise, the EKF is a very good observer for the systems that present any kind of noise. It will exploit the noise in order to estimate the quantity. The process of observation of the EKF is given in two stages, prediction and filtering. The prediction stage is aimed to obtain the next predicted states and predicted state-error covariance, while in the filtering stage, the next estimated states is obtained as the sum of the next predicted states and a correction term.

## 6. Conclusions

All the closed-loop observers are classified as deterministic observers, they can be easily corrupted by measuring noise and require parameter adaptation algorithms. The Kalman filter observer has high convergence rate and good disturbance rejection, which can take into account the model uncertainties, random disturbances, computational inaccuracies and measurement errors. These properties are the advantages of extended Kalman filters over other estimation methods. For these reasons, it had wide application in sensorless control in spite of its computational complexity. For non-linear problems Kalman filtering can overcome this difficulty by using a linearized approximation, where, the stochastic continuous time system must be expressed in the discrete form in order to fit with the structure of the EKF. The process of observation of the EKF is given in two stages, prediction and filtering. The prediction stage is aimed to obtain the next predicted states and predicted state-error covariance, while in the filtering stage, the next estimated states is obtained as the sum of the next predicted states and a correction term.

However, the high degree of complexity of EKF structure and the high system orders cause a higher computational requirement (the sampling time). Thus, additional challenges and problems are introduced, such as the reduction of dynamic performance and the increase of harmonics. Nevertheless, the development of new processors technology (DSPs and FPGA) solves this problem due to the powerful calculations processing.

Recently, different works have been conducted to improve the effectiveness and the performance of the sensorless EKF for IM drive control. A bi-input EKF estimator, which deals with the estimation of the whole state of the machine together with stator and rotor resistances is presented. Another multi-model EKFs are proposed in order to improve EKF performance under different noise conditions. Then, a Kalman filter estimator has been designed for DTC controlled induction motor drives.

## Acknowledgements

The authors would like to thank the managers and the members of the “Laboratoire d’électronique de puissance et commande (EPC) de l’Ecole Mohammadia d’Ingénieurs (EMI)” for their precious remarks and suggestions.

## Notations and symbols

| | |
|---|---|
| $i_{s\alpha}, i_{s\beta}$ | stator current components in $\alpha - \beta$ reference frame |
| $\phi_{r\alpha}, \phi_{r\beta}$ | rotor flux components in $\alpha - \beta$ reference frame |
| $u_{s\alpha}, u_{s\beta}$ | stator voltage components in $\alpha - \beta$ reference frame |
| $R_s, R_r$ | stator and rotor resistances |
| $L_s, L_r, L_m$ | stator, rotor and mutual inductances |
| $T_r$ | rotor time constant |
| $T_{em}$ | electromagnetic torque |
| $T_L$ | load torque |
| $\omega_r$ | electrical speed |
| $\Omega_r$ | mechanical speed |
| $\sigma$ | Blondel’s coefficient |
| $p$ | pole pair number |
| $J$ | inertia moment |
| $f$ | friction coefficient |
| $A, B, C$ | control, input and output matrices of the induction motor model |

## Abbreviations

| | |
|---|---|
| EKF | extended Kalman filter |
| ALO | adaptive Luenberger observer |
| IM | induction motor |
| PI | proportional-integral |
| *d-q* | direct-quadrature |
| MatLab | matrix laboratory |
| DSP | digital signal processor |
| FPGA | field-programmable gate array |
| DTC | direct torque control |
| AC | alternating-current |

# A. Appendix

| Parameter [a] | Rated value |
|---|---|
| Power | 3 $kW$ |
| Voltage | 380 $V$ |
| Frequency | 50 $Hz$ |
| Pair pole | 2 |
| Rated speed | 1440 $rpm$ |
| Stator resistance | 2.20 Ω |
| Rotor resistance | 2.68 Ω |
| Stator inductance | 0.229 $H$ |
| Rotor inductance | 0.229 $H$ |
| Mutual inductance | 0.217 $H$ |
| Moment of inertia | 0.047 $kg.m^2$ |
| Viscous friction coefficient | 0.004 $N.s/rad$ |

[a]*Used induction motor rated parameters.*

**Table A1.**
*The parameters of the used induction motor in simulation.*

# A. Computer program

```
function[ sys,x0] =EKF(t,x,u,flag)
global a1 a2 a3 a4 a5 b1;
global F C K R Q P Ts B n p;
if flag==0
% Machine parameters——————————————
Rs=2.2;Rr=2.68;M=0.217;Ls=0.229;Lr=0.229;p=2;
% Initiating the state error covariance matrix————————
P=eye(5);
% State noise covariance matrix——————————
Q=diag([ 1e-6 1e-6 1e-6 1e-6 1e6] );
% Measure noise covariance matrix——————————
R=diag([ 1e6 1e6] );
% Sampling period
Ts=1e-5;
% Defining A and B matrices——————————
Tr=Lr/Rr;
Sigma=1-M^2/(Ls* Lr);
a1=-(Rs/(Sigma* Ls)+(1-Sigma)/(Sigma* Tr));
a2=M/(Sigma* Ls* Lr* Tr);
a3=M/(Sigma* Ls* Lr);
a4=M/Tr;
a5=-1/Tr;
b1=1/(Sigma* Ls);
% Input Matrix——————————————
B=[ b1 0;0 b1;0 0;0 0;0 0] ;
```

```
% Measure Matrix------------------------------------------
C=[ 1 0 0 0 0;0 1 0 0 0] ;
% Loop ------------------------------------------
n=0;
x0=[ 0 0 0 0 0] ;
sys=[ 0,5,5,4,0,0] ;
elseif flag==2
n=n+1;
U=[ u(1);u(2)] ;
Y=[ u(3);u(4)] ;
A=[a1       0        a2          a3*x(5)      a3*x(4)
   0        a1       -a3*x(5)    a2           -a3*x(3)
   a4       0        a5          -x(5)        -x(4)
   0        a4       x(5)        a5           x(3)
   0        0        0           0            0];
F=eye(5)+Ts*A;
G=Ts*B;
% State prediction------------------------------------------
x_1=[ F(1:4,1:4)*x(1:4);x(5)] +G*U;
% Covariance prediction------------------------------------------
P_1=F*P*F'+Q;
% Kalman gain matrix------------------------------------------
K=P_1*C'/(C*P_1*C'+R);
% State estimation------------------------------------------
x=x_1+K*(Y-C*x_1);
% State error covariance estimation------------------------------------------
P=P_1-K*C*P_1;
    sys=x;
elseif flag==3
    sys=x;
elseif flag==9
    sys=[ ];
end
```

## Author details

Yassine Zahraoui*† and Mohamed Akherraz†
Mohamed 5 University, Mohammadia School of Engineering, Agdal-Rabat, Morocco

*Address all correspondence to: zahraoui.yassin@gmail.com

† These authors did not contribute equally.

## References

[1] Brown RG. Introduction to Random Ssignal Analysis and Kalman Filtering. Vol. 8. New York: Wiley; 1983

[2] Lefebvre T, Bruyninckx H, Schutter J. Nonlinear Kalman Filtering for Force-Controlled Robot Tasks. Vol 19. Berlin: Springer; 2005

[3] Brown RG, Hwang PYC. Introduction to Random Signals and Applied Kalman Filtering. Vo. 3. New York: Wiley; 1992

[4] Kron G. Generalized theory of electrical machinery. Transactions of the American Institute of Electrical Engineers. 1930;**49**(2):666-683

[5] Zahraoui Y, Fahassa C, Akherraz M, Bennassar A. Sensorless vector control of induction motor using an EKF and SVPWM algorithm. In: 5th International Conference on Multimedia Computing and Systems (ICMCS). Marrakech: IEEE; 2016. pp. 588-593. DOI: 10.1109/ICMCS.2016.7905584

[6] Ananthasayanam MR. A reference recursive recipe for tuning the statistics of the Kalman filter. In: de Oliveira Serra GL, editor. Kalman Filters—Theory for Advanced Applications. Rijeka: IntechOpen; 2018. DOI: 10.5772/intechopen.71961

[7] Zahraoui Y, Akherraz M, Fahassa C, Elbadaoui S. Robust control of sensorless sliding mode controlled induction motor drive facing a large scale rotor resistance variation. In: Proceedings of the 4th International Conference on Smart City Applications, SCA '19 (Casablanca, Morocco), Association for Computing Machinery, Oct. 2019; 2020. pp. 1-6

[8] Tahar B, Bousmaha B, Ismail B, Houcine B. Speed sensorless field-oriented control of induction motor with fuzzy luenberger observer. Electrotehnica, Electronica, Automatica. 2018;**66**(4):22

Chapter 6

# Data Processing Using Artificial Neural Networks

*Wesam Salah Alaloul and Abdul Hannan Qureshi*

## Abstract

The artificial neural network (ANN) is a machine learning (ML) methodology that evolved and developed from the scheme of imitating the human brain. Artificial intelligence (AI) pyramid illustrates the evolution of ML approach to ANN and leading to deep learning (DL). Nowadays, researchers are very much attracted to DL processes due to its ability to overcome the selectivity-invariance problem. In this chapter, ANN has been explained by discussing the network topology and development parameters (number of nodes, number of hidden layers, learning rules and activated function). The basic concept of node and neutron has been explained, with the help of diagrams, leading to the ANN model and its operation. All the topics have been discussed in such a scheme to give the reader the basic concept and clarity in a sequential way from ANN perceptron model to deep learning models and underlying types.

**Keywords:** ANN, artificial neural network, node, network training, gradient descent, deep learning

## 1. Introduction

Artificial Intelligence (AI) is the knowledge domain that targets the development of computer systems to solve problems by giving them cognitive powers for performing tasks that usually require human intelligence. Hence, simulation of human intelligence, with computer programing and technologies, is the main objective of AI. Whereas, machine learning is one of the branches of AI, in which computer systems are programmed based on the data and type of input. Machine learning (ML) gives the capability to AI for solving problems based on available data. Likewise, artificial neural network (ANN) is an evolved method of ML algorithms, developed on a concept of imitating the human brain [1–3].

A single neuron is considered as a cell, processing electrochemical signals or nerve impulses, and the human brain is a complicated network of neurons that transfers information, with the help of various interlinked neurons. ANN models are considered as most popular among AI models because of their architecture, which is the collection of neurons linked with other neurons in various layers. ANN is non-linear and complex systems of neurons and neuron is a mathematical unit [4].

Literature depicts that ML, ANN and deep learning (DL) falls under the pyramid of AI and shown in **Figure 1**. Under ANN, DL has gained much importance among researchers. DL is a complex network set of ANN with various layers of processing, which improves the results by developing high levels of insight. DL methodologies

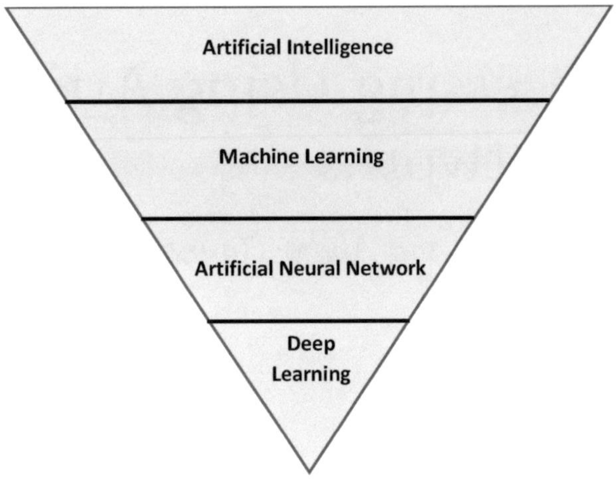


**Figure 1.**
*AI pyramid.*

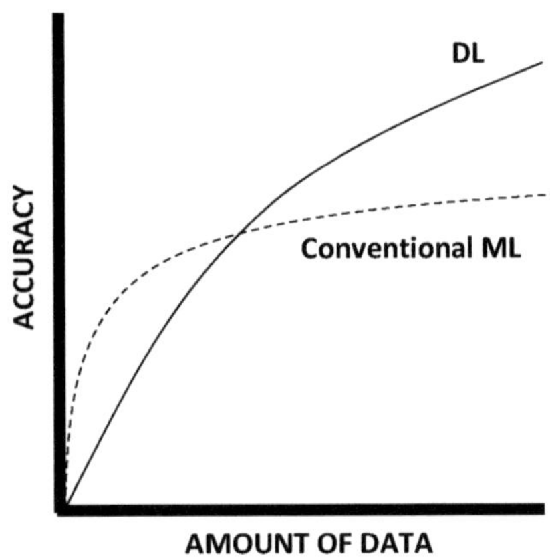


**Figure 2.**
*Comparison between DL and conventional ML.*

are popular due to their computational powers and handling of large data sets, and this makes them more attractive than conventional methods.

Past studies illustrated the comparison between DL and conventional ML methods for effective outputs, with the help of graphical representation, as shown in **Figure 2**. **Figure 2** illustrates the behaviour of curves, for DL and conventional ML, by comparing the accuracy of results (outputs) against the amount of data (input). The graph shows that the result accuracy of conventional ML methods is better for limited data, but it decreases as the amount of data is increased. Instead, the result accuracy of DL improves for large data sets, due to the presence of a vast neural network than conventional ML, hence, making DL more famous. DL is usually used for complicated tasks, such as image classification, image recognition, and handwriting identification [1, 3].

## 2. History of ANN

The origins of all the work on ANN are in neurobiological studies that date back to about a century ago. A brief overview of evolution in ANN and significant milestones are shown in the timeline, as shown in **Figures 3** and **4**.

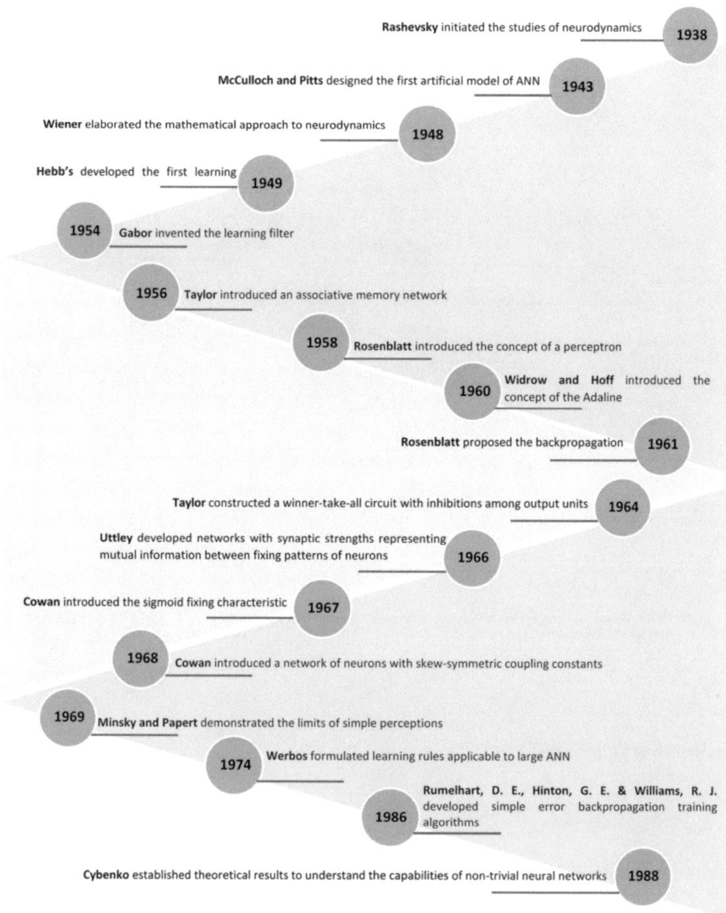


**Figure 3.**
*ANN evolution timeline (1938–1988).*

Literature depicts that, in the 1980s, very few researchers were working on deep NNs, and it gained popularity in the early 1990s. Since then, a large number of research articles have been published on applications of ANN and this journey is on-going. The few significant milestones, after 1990, regarding ANN evolution is shown in **Figure 4** [5–10].

## 3. Basic architecture of ANN

The architecture of ANN is stimulated by the framework of biological neurons, like in the human brain. The human brain is the composition of a vast number of the interlinked neurons forming a network. A neuron is like a cell, and each neuron executes a simple task, i.e., response to an input signal. Likewise, the ANN is a framework of interlinked nodes, similar to neurons, forming a network model. Hence in ANN, several artificial neurons are interlinked and become a robust computer-based tool that can handle large amounts of data to execute enormously simultaneous calculations using input data. ANN operations are not based on explicit rules and outputs are generated by trial and error procedures through sequential computations. The ANN is also classified as 'connectionism' because the given data is not conceded from neuron to neuron, but it is encoded in the complicated interconnected network of neurons, unlike the traditional computers [2, 11, 12].

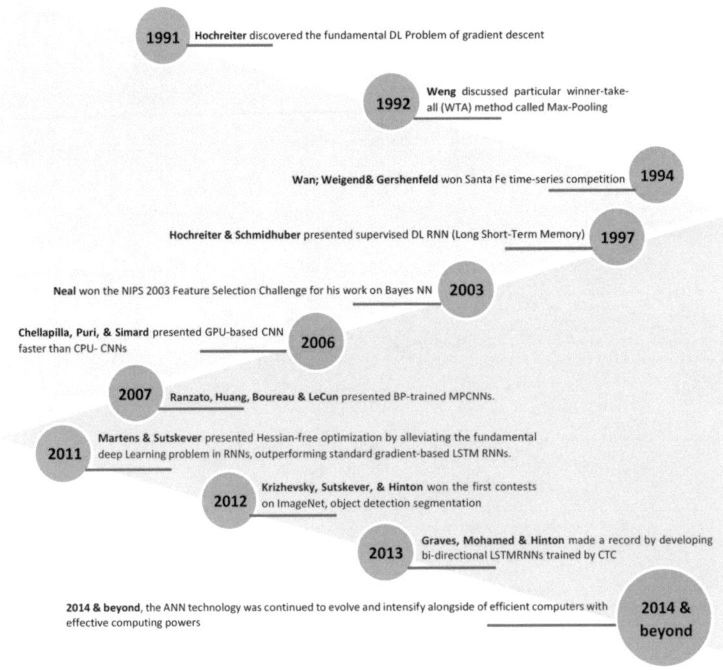


**Figure 4.**
*ANN evolution timeline (after 1988).*

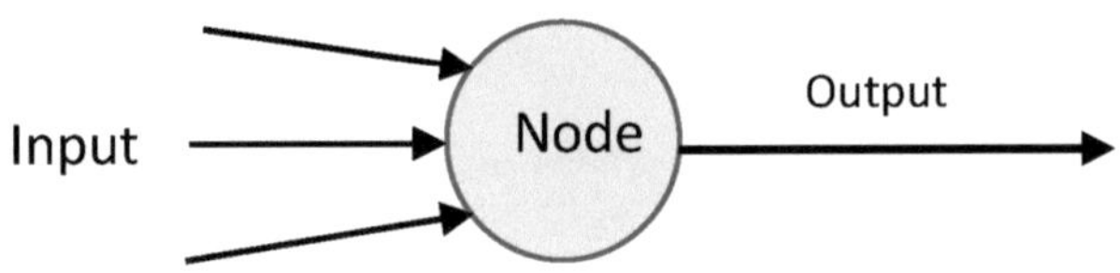


**Figure 5.**
*Basic node model.*

To comprehend the basic structure of ANN, firstly, the understanding of 'node' is necessary. The generic model for a node is shown in **Figure 5**.

Each node receives various inputs through connections and transfers it to adjacent nodes. **Figure 6** represents the general model of ANN, which is stimulated by a biological neuron.

The nodes are arranged and organised into linear arrays known as layers. **Figure 6** shows that there are three layers in ANN called the input layer, the output layer and the hidden layer.

In the input layer $X_1, X_2, X_3, \ldots X_n$ signifies several inputs to the network. Whereas, $W_1, W_2, W_3, \ldots W_n$ are known as connection weights, which shows the strength of a particular node. In ANN, weights are considered as the most significant factors as these are numerical parameters that determine the effect of neurons to each other and also impact the output, by converting the input.

In the ANN, the processing part is performed in the hidden layer. The hidden layer executes two operational functions, i.e., summation function and transfer function, also known as an activation function. The summation function is the first step, and in this part, each input ($X_i$) to ANN is multiplied by its respective weight ($W_i$) and then, the products $W_i.X_i$ is cumulated into the summation function $\xi = \Sigma W_i.X_i$. 'B' is a bias value; this parameter is used to regulate the output of the neuron in association with the weighted sum of the inputs. This process is denoted as Eq. (1):

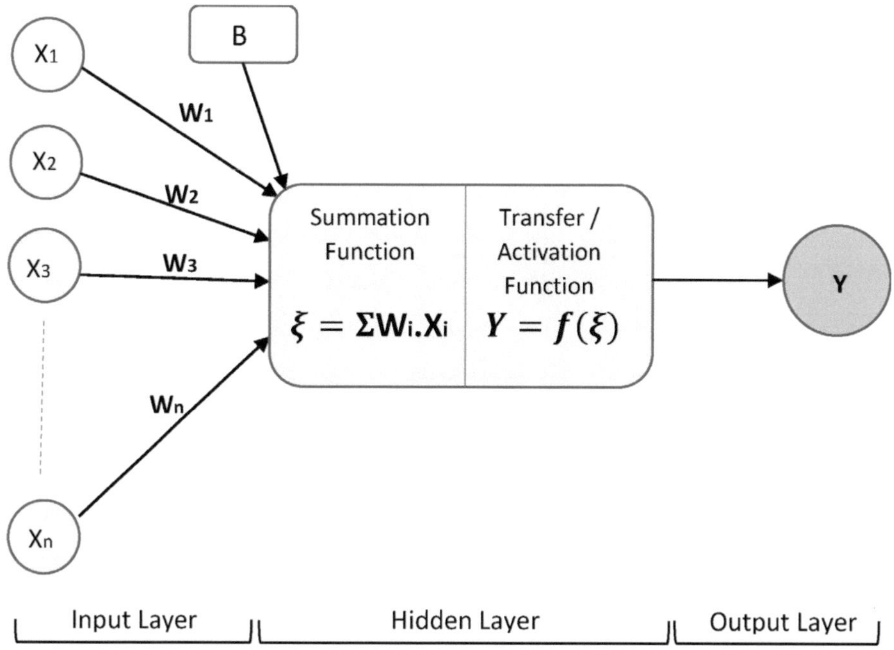


**Figure 6.**
*Generic ANN model.*

$$Output = \mathbf{\Sigma}(Weights \times Inputs) + Bias \tag{1}$$

The activation function is the second step; which converts the input signal, received from the summation function module and transformed it to an output of a node for an ANN model [1–3, 12, 13].

Generally speaking, each ANN has three main components, i.e., node character, network topology and the learning rules. The node character controls the processing of signals by determining the associated number of inputs and outputs, the associated weight for each input and output and the activation function, for each node. Learning rules establish the initiation and adjustment of weights. Whereas, the network topology defines the ways the nodes will be connected and organised (details are discussed in Section 3.2). The operation of the ANN model is computing the output of all the neurons, which is an entirely deterministic calculation [1, 2].

### 3.1 The activation function

An activation function is a mathematical function. In simple words, it receives the output of the summation function as an input and converts that into the final output of a node with the help of ANN processing.

There are different types of activation functions, but non-linear functions are more popular than the linear function. A linear function is just a polynomial of one degree, and it is considered as single-layer ANN model has less power and limited complexity to process complicated data. Therefore, non-linear activation functions are mostly included in designing of ANN models for solving complex problems and this unique quality makes ANN true universal function approximators.

The activation function uses the value $\xi = \Sigma W_i.X_i$ as an input for processing and controlling the input $X_i$ for activation of the neuron. The most commonly known activation functions [1, 12–15] are shown in **Table 1**.

| Transfer functions | Graphical presentation | Numerical equation | Remarks |
|---|---|---|---|
| Linear | 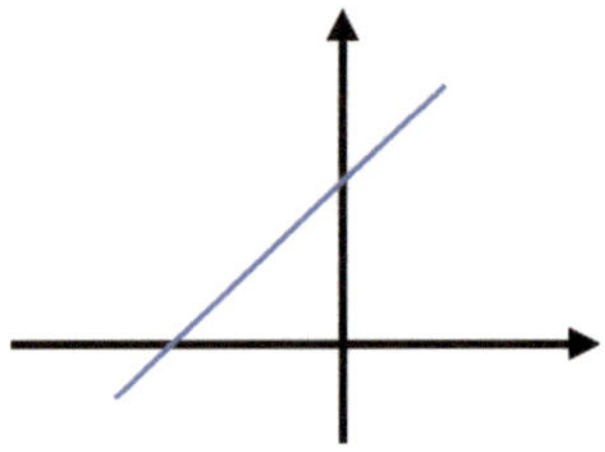 | $Y = f(\xi) = \xi$ | Output = Input. Range $(-\infty, +\infty)$ |
| Unit step | 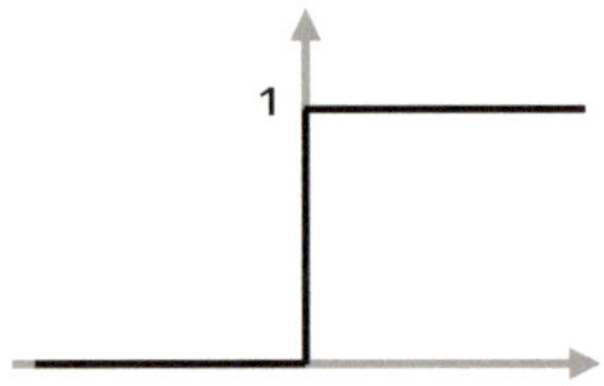 | $f(\xi) = 0$ if $\xi < 0$ $f(\xi) = 1$ If $\xi \geq 0$ | Useful for binary schemes. Range (0,1) |
| Rectified linear unit (ReLU) | 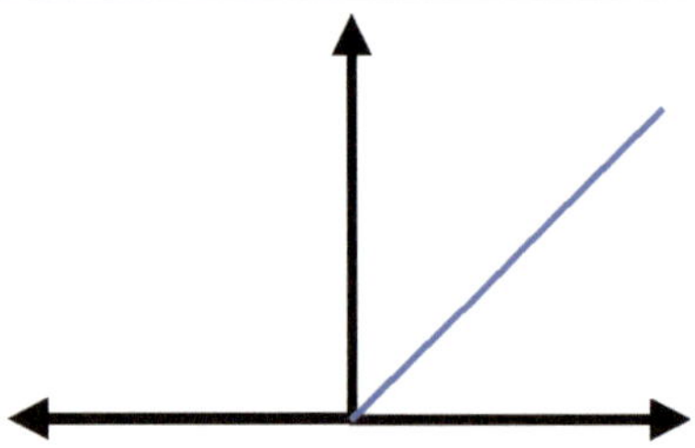 | $f(\xi) = 0$ if $\xi < 0$ $f(\xi) = \xi$ If $\xi \geq 0$ | Most popular activation function since 2015. Range $(0, \infty)$ |
| Sigmoid | 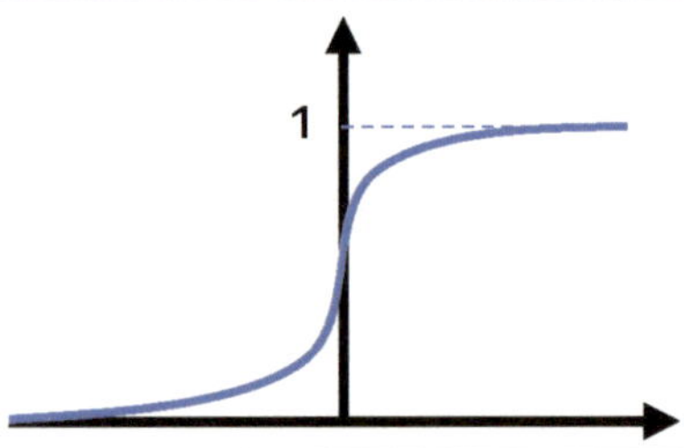 | $f(\xi) = \frac{1}{1+e^{\xi}}$ | Commonly used function. Range (0, 1) |
| Gaussian | 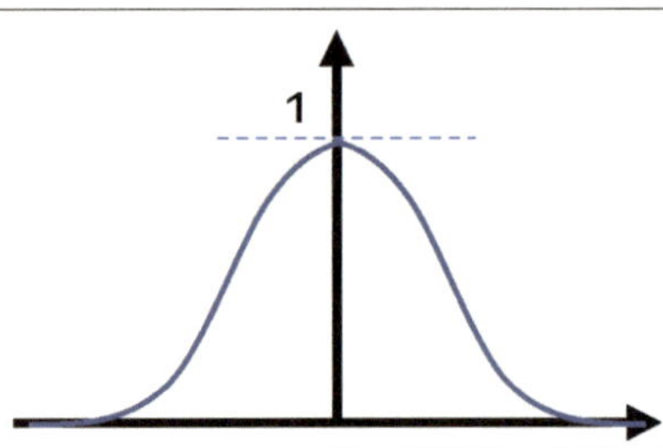 | $f(\xi) = e^{-(\xi)^2}$ | Named after the mathematician Carl Friedrich Gauss Range (0,1] |
| Hyperbolic tangent | 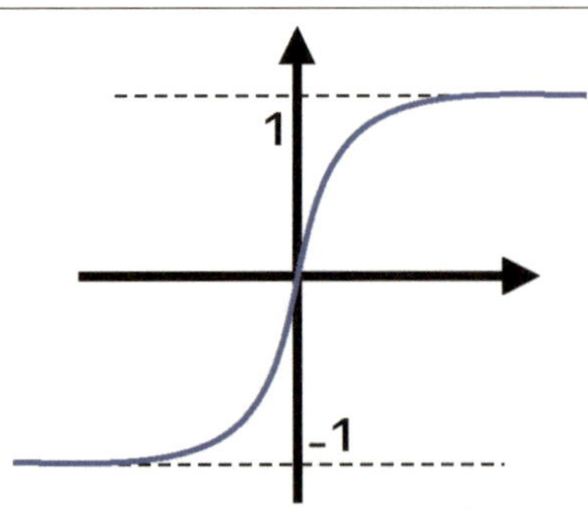 | $f(\xi) = \frac{2}{1+e^{-2\xi}}$ - 1 | Alternative to sigmoid function. Range $(-1, 1)$ |

**Table 1.**
*Activation functions.*

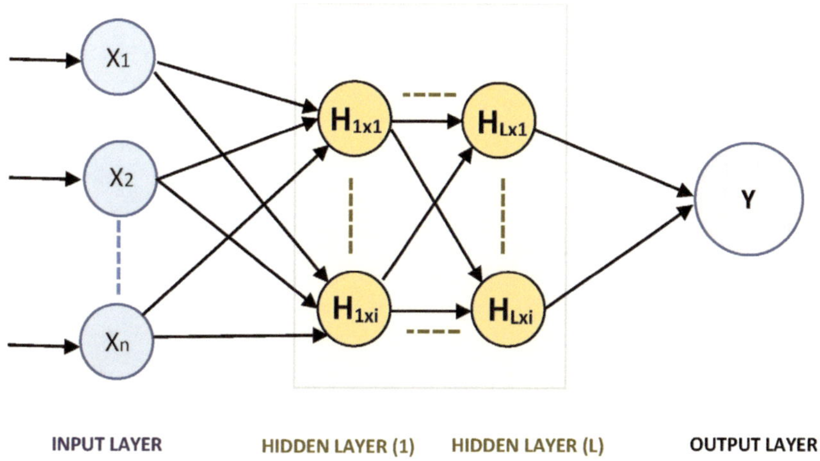


**Figure 7.**
*Conceptual model for ANN topology.*

## 3.2 Network topology

The nodes are arranged and organised into linear arrays known as layers. The interconnecting network model, between the nodes of ANN, with each other, is called the topology (or architecture). ANN is composed of input layers, hidden layers and output layers, as already discussed in **Figure 6**. Also, the hidden layers can be from none to numerous, based on the model-complexity. Each layer is a combination of many nodes, and these nodes, based on some properties, can be grouped in layers. A single-layer ANN, with a single output, is known as Perceptron. A conceptual model for layers and ANN topology is shown in **Figure 7**. **Figure 7** shows n number of data entries in the input layer as $X_1$, $X_2$, …. $X_n$. Also, it can be seen that there is L number of hidden layers in the ANN model. Whereas, there are i number of nodes in each hidden layer. The notations $1 \times 1$, $1 \times i$, $L \times 1$ and $L \times i$, on each node giving its information, expressing 'L' as (hidden) layer number, i.e., from 1 to L and 'i' as node number, i.e., from 1 to i. Y is the output for the mentioned ANN model.

Designing of network topology is based on following factors; (1) the number of nodes in each layer, (2) the number of layers in the network and (3) the connected path among the nodes [1, 2, 12].

### *3.2.1 Perceptron and multi-layer architectures*

A single-layered ANN, with a single output, is known as the perceptron. The perceptron mostly uses the step function, in which, if the computed sum of the inputs transcends a threshold point, the output is 1; otherwise, it is 0.

Multi-layer perceptrons (MLPs) are the most commonly used architecture for ANN. Composition of MLPs contains layers of neurons with an input layer, an output layer, and the hidden layer (at least one). The layers of the perceptron are interlinked with each other by developing a multi-layered architecture, and this makes the model essentially complex for the ANN processing. The MLP terminology is originated from perceptron neural networks, but its problem-solving capabilities makes it unique [1, 14].

## 3.3 Connection types between nodes

The connections between nodes of ANN are classified into two categories: (1) the feedforward network, and (2) the feedback network or recurrent network.

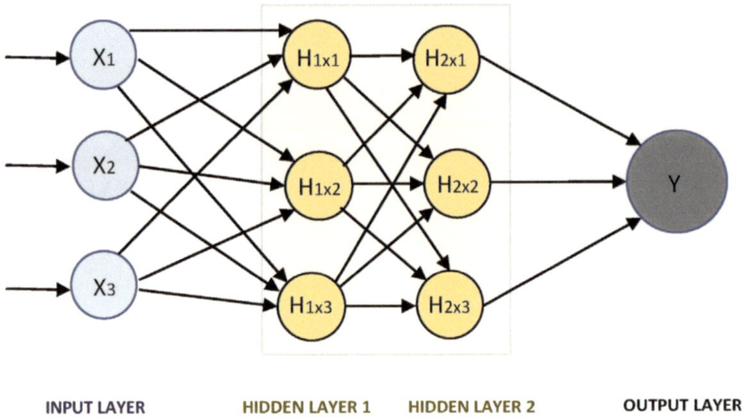


**Figure 8.**
*Feedforward network connection.*

### *3.3.1 Feedforward networks*

Feedforward network is a one-way connection having no loop backwards. They are static in nature as their signal travels in one way only. **Figure 8** is a model example of feedforward networks.

### *3.3.2 Feedback networks*

In feedback network, nodes have backward connected loops, and in these connections, the output of the nodes can be the input to the same level or previous nodes. Unlike the feedforward network, the feedback networks are dynamic. In feedback networks, signals are transmitted in forward as well as in backward directions [16]. Feedback process occurs when the output (partial or full) is channelled back into the input of a network as part of a repeated cause-and-effect process [17]. In the feedback network, a single input generates a series of outputs cycles until it reaches an equilibrium point. Equilibrium point refers to minimum error, i.e., for each predicted output if the error is enormous then, the output is routed back, and parameters (weights and biases) are modified until the error becomes minimum [18]. **Figure 9** shows the ANN model for feedback network

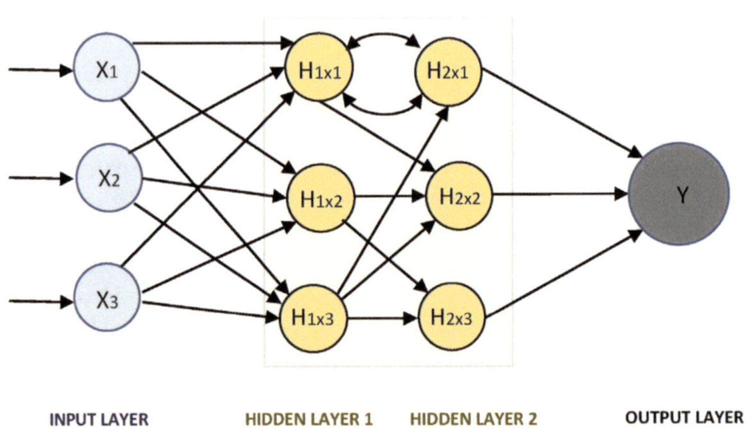


**Figure 9.**
*Feedback network connection.*

connections. It can be observed that node H2x1 is sending the information back to node H1x1 and the cycle goes on until the output will reach an equilibrium state, i.e., with minimum error. In a feedback network, there exists at least one interconnected path that drives it back to the starting neuron. It may cause a delay in specific time units, and this interconnected path is called a cycle [1, 2, 12]. This process will be better understood, after going through the next section.

## 4. Training of ANN (learning process)

The training of the ANN is accomplished through a learning process. While in the training process, weights are modified for attaining required results. In the training process, some sample data is processed to the network and weights are modified to attain better approximation of the desired output.

The learning process is mostly classified into two categories: (1) supervised learning, and (2) unsupervised learning.

### 4.1 Supervised learning

In supervised learning, a training set is presented to the model. The training set constitutes of input examples and corresponding target outputs. The inputs are noted for the response of the network, and the weights between with networks are adjusted for error reduction, for the attainment of the desired output. The network follows successive iterations during this process until the computed result converges to the correct one. Construction of the training set requires special consideration. A training set is considered an ideal one, and it should be giving a better representation of the underlying model. Otherwise, a reliable model with desirable results cannot be achieved with an unrepresentative training set.

In the supervised learning process, the networks are trained first before its operation in a model for predictive outputs. Significantly, when the network starts computing the intended outputs with the series of inputs, with fixed weights, then the ANN model can be set for the required operation. Few of the well-known algorithms with a supervised learning method are the Adaline (used for binary data), the Perceptron (used for continuous data), and the Madaline (developed from the Adaline).

#### *4.1.1 Reinforcement learning*

Reinforcement learning is a particular case scenario of supervised learning. It is, when the external environment only checks for the information for acceptance and rejection, instead of indicating the correct output. In this process, the well-performing and the most active neuron connections for the input are strengthened over successive iterations. Few of the renown algorithms of reinforcement learning are the Boltzmann machine, the learning vector quantisation, and Hopfield networks.

Supervised ANN models have many applications for image classification, plant control, forecasting, prediction, robotics, ECG signals classification and many more [19–21].

### 4.2 Unsupervised learning

Unsupervised learning does not follow a training set or a targeted output approach. Instead, it trails the input data pattern of the underlying model. In this process, the ANN model adjusts its weights, against the supplied inputs, thus producing outputs similar to inputs. The model, without any outer support, recognises the patterns and differences in the inputs. In this process, the clusters are formed, each cluster consists of a group of several weights, in such a way that related input path results in a similar output. If any new pattern is detected during the iteration process, it is classified as a new cluster.

Autoencoders, Hebbian Learning, Deep Belief Nets, Self-Organising Map, Generative Adversarial Networks, and Algebraic Reconstruction Technique (ART) are the few most renown algorithms for unsupervised learning. Unsupervised ANN models are used in diagnosing diseases, image segmentation and many more. Unsupervised algorithms have become very useful and powerful tools in segmentation of magnetic resonance images for detection of anomalies in the body systems [1, 2, 4, 12, 14, 22–24].

## 5. Mapping by ANNs

The primary reason for ANN popularity is due to approximated data output. There are five main steps for the approximation function in the ANN model, as given below.

### 5.1 Data pre-processing

In data pre-processing, the appropriate predictors are selected as inputs before processing to a network for mapping. There are three general processes in data pre-processing, mentioned as follows:

a. *Standardising*: The input values are rescaled to a uniformed scale.

b. *Normalising*: It normalises a vector to have unity variance and zero mean value.

c. *Principal component analysis*: This process replaces the groups of related variables by new unrelated variables by detecting linear dependencies between them.

### 5.2 Selection of network architecture

A network architecture comprises several hidden neurons, the number of hidden layers, the flow of data, the way neurons are interconnected, and specific transfer functions. Recurrent neural networks, multi-layer perceptron (MLP), probabilistic neural networks, radial basis function networks, generalised regression neural networks and time-delay neural networks are the few of the renown architectures.

### 5.3 Network training

About function mapping, the training process is known as the calibration of the network through input and out pairs. During the training process, ANN might suffer from the overfitting and underfitting. The overall performance of the network decreases because of these two mentioned factors. This unfitting of the network, during the training process, can be managed by increasing the number of epochs, but it may result in network overfitting if the number of epochs is more significant than a required number. Epoch is defined as a process of providing one pass or iteration of input through the network and modifying the weights. The optimal number of epochs can be determined by the comparison of training error and model testing procedure.

### 5.4 Simulation

Simulation is the ultimate goal of applying ANN networks. It is the representation of predicted output data for an ANN model.

### 5.5 Post-processing

There are three types of sets in which sample data is distributed: (i) the training set, (ii) the validation set, and (iii) the testing set. The training set is used to train the ANN model; it is a set of sample data that is used to modify or adjust the weights in the ANN to produce the desired outcome. The validation set is used to inform the ANN when training is to be terminated (when the minimum error point is achieved). The test set provides an entirely independent way of examining the precision of the ANN. The test set is a set of sample data that is used for the evaluation of the ANN model. A rule of thumb for this random split regarding percentage is 70, 15, 15%, respectively [3, 12, 14].

The post-processing comprises of all the tests, which are applied on a specific network for the validation of results, also, to analyse, describe, and to improve its final performance. The comparison of results is achieved by using three different statistics. The first one is the root-mean-square error (RMSE), and it is described in Eq. (2):

$$RMSE = \sqrt{\frac{\sum_{i}^{n} (obs_i - est_i)^2}{n}} \tag{2}$$

The second statistical factor is percentage volume error (%VE), which is the measuring of the absolute relative bias error of estimated values. It is formulated as Eq. (3):

$$\%VE = \frac{\sum_{i=1}^{n} \left(\frac{obs_i - est_i}{obs_i}\right)}{n} \tag{3}$$

whereas, $est_i$ = ith estimated variable, $obs_i$ = ith observed data, and n = number of observed values.

The third statistical factor is the correlation, and it is used in the measuring of the linear correlation coefficient between the predicted and observed data.

In case of unsatisfactory results in the post-processing, modification can be made in the following: (1) weights and biases, (2) number of hidden neurons, (3) transfer functions, and (4) number of hidden layers [4, 25].

## 6. Gradient descent

The term 'gradient descent' is a combination of two words the 'gradient', which means a slope and the 'descent', which means to incline. Therefore, with gradient descent, the slope of gradients is descended to find the lowest point with the smallest error. It is an iterative process until the correction of the error in the ANN learning model. It is defined as during the backpropagation in the ANN model, the process of iteration keeps updating biases and weights with the error times derivative of the activation function. The steepest descent step size is substituted by a similar size from the previous step.

A gradient is the derivative of the activation function, as shown in **Figure 10**.

The primary purpose of using gradient descent is to find the overall cost minimum at each step, with the lowest error. Also, at this point, model predictions are more reliable because of upright fit data. Evaluation of slope can be done with the help of **Figure 11**, and Eq. (4) can be derived.

$$\Delta x_i = -\alpha \frac{dy}{dx_i} \tag{4}$$

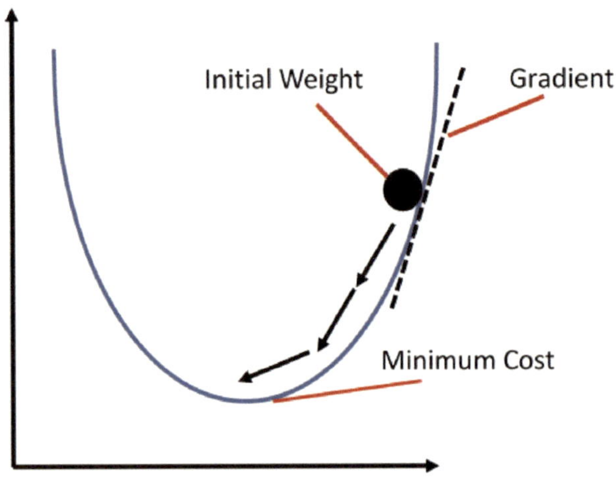


**Figure 10.**
*Gradient descent.*

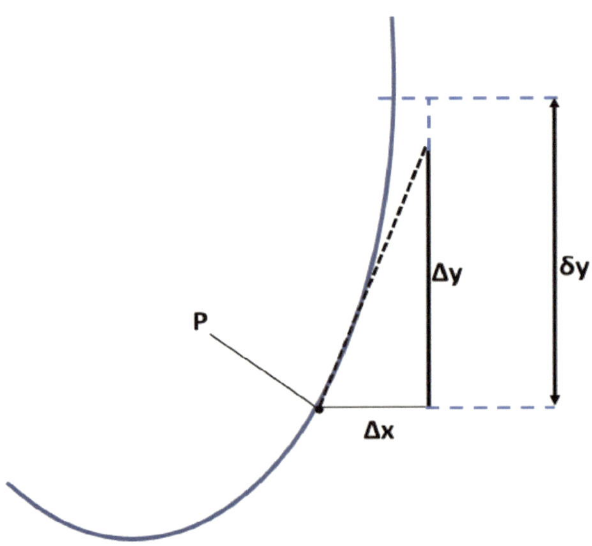


**Figure 11.**
*Slope computation.*

whereas, $\alpha$ = learning rate and $dy/dx_i$, also known as the partial derivative of y with respect to $x_i$. For gradient descent, this equation can be used for each variable when $\delta y < 0$ ($\delta$ is a partial derivative).

Gradient descent can be achieved either for the stochastic or full batch. In stochastic, gradient descent performs calculation for gradient by taking a single sample. Whereas, in full batch, the gradient is calculated for the full training dataset. One of the advantages of stochastic gradient descent is the fast calculation of gradients [1, 13, 23].

### 6.1 Training algorithm by delta rule

The biases and weights are the parameters of the network that are required to be adjusted before operating an ANN. These parameters can be modified by using either supervised or unsupervised approach for any ANN model. For training purpose, the supervised learning process is generally considered for determining biases and weights of an ANN network. The supervised training process of an ANN network could be attained by using delta rule. The delta rule is expressed as $W_{ij}$ with the help Eqs. (5)–(7), as shown:

$$W_{ij}^{new(L)} = W_{ij}^{old(L)} + \alpha\left(-\frac{\partial_{e_p}}{\partial_{W_{ij}^{(L)}}}\right) \quad (5)$$

$$E = \frac{1}{n}\sum\nolimits_{p=1}^{n} e_p \quad (6)$$

$$e_p = \left(t_p - y_p\right) \quad (7)$$

whereas, n = the number of pairs of data, W = the weight of the link between the ith neuron to the jth neuron in the Lth layer, E = the average error of estimation, $t_p$ = target output, $y_p$ = simulated output, $\alpha$ = learning rate, the value of which is selected between 0 and 1 experimentally.

The backpropagation algorithm is mostly used for the application of delta rule for the training process of an ANN. The mathematical expression of delta rule is changed to computational relation because of the backpropagation algorithm, which can be applied through an iterative process. This process provides a way to the gradient for determining of the minimum error function, and it is efficiently calculated by using the chain rule of differentiation provided by the backpropagation algorithm. This characteristic makes this process to also be known as the generalised delta rule. In this algorithm, during each iteration, the network weights are shifted along with the negative of the gradient in the steepest descent direction of the performance function (epoch). For a certain weight in the Lth hidden layer, the chain rule gives Eq. (8):

$$\frac{\partial_{e_p}}{\partial_{W_{ij}}^{(L)}} = \frac{\partial_{e_p}}{\partial I_{pj}^{L}} \cdot \frac{\partial I_{pj}^{(L)}}{\partial w_{ij}^{(L)}} \quad (8)$$

This algorithm keeps the iterations continued until the expected output of network training is achieved. The basis for stopping the training process may be the minimum target value of performance function, the number of epochs and run time of the process; this is known as stopped training. The above mentioned equations lead to the following weight calculating Eqs. (9) and (10):

For the last layer

$$W_{ij}^{new} = W_{ij}^{old} + \alpha\delta_{pj}y_{pi}^{(L-1)} \tag{9}$$

For the hidden layer

$$W_{ij}^{new} = W_{ij}^{old} + \alpha\delta_{pj}^{(L)}y_{pi}^{(L-1)} \tag{10}$$

Following this procedure of training, based on the specific input vectors using the final derived weights and biases, the ANN model will be operated on sample data for initiation of simulation for the related outputs. The ANN training can be achieved either by batch training or incremental training. During the batch training process, the adjustment of biases and weights is attained after the presentation of all the inputs and targets. Whereas, during the incremental training, the adjustment of biases and weights is attained just after the presentation of individual input. In training, the process affects network performance. In the case of the low learning rate, the time required for learning the synaptic weights will be extremely long. On the other hand, if the set learning rate will be too high, this will tend the algorithm to oscillate, and the trained network performance will be reduced because the weight changes are too drastic. Therefore, the learning rate controls the convergence of the algorithm. These weight modifications can be applied after each pattern is completed, and these computed weight changes can be summed up to be applied to the network weights, as shown in Eq. (11):

$$\Delta w_{ij}^{l} = \sum_{p=1}^{n} \Delta w_{pij}^{L} \tag{11}$$

Usually, in dynamic networks, the inputs and targets are shown in sequence. In the adaptive learning process, the recent data, that is perceived before the time of simulation is considered as necessary as compared to all the data [4, 14, 26].

## 7. Deep learning

In the field of AI, deep learning (DL) has gained much popularity and trending for investigation domains. One of the foremost shortcomings of conventional machine learning is their inability to solve the selectivity-invariance problem, and because of this drawback, these methods have limited capability of data processing in their real state. Selectivity-invariance enables the model for the selection of those parameters that comprise of more information and disregard parameters with less information. This characteristic of DL, i.e., ability to overcome the selectivity-invariance dilemma, makes it more likeable among researchers and motivate them to the advancement of machine learning using the DL approach.

The architecture of DL is composed of various layers of trainable parameters, and this helps DL-based algorithms for excellent performance in machine learning and AI applications. DL algorithm is Deep Neural Networks (DNNs), and they usually use backpropagation optimised algorithms for end-to-end training. DNNs capability of selectivity-invariance extracts the compound features through successive layers of neurons equipped with differentiable, non-linear activation functions, and this provides a suitable platform for the backpropagation algorithm. A generic architectural model of DNNs is shown in **Figure 12**.

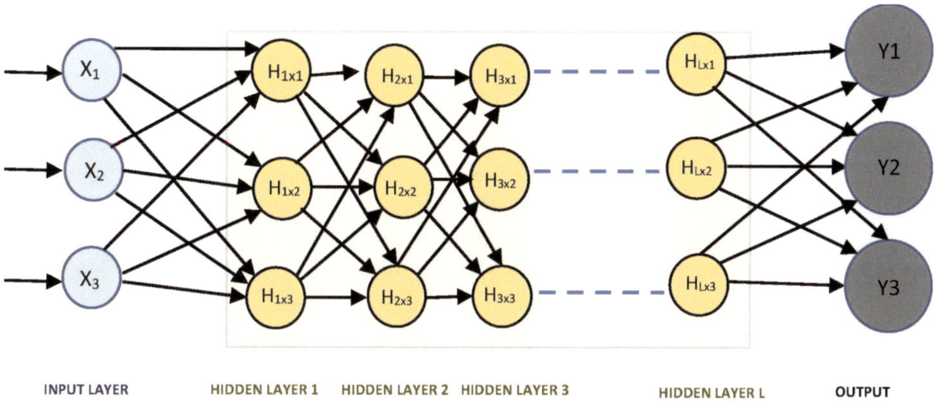


**Figure 12.**
*DNNs generic model.*

**Figure 12** depicts a DNNs model with numerous hidden layers. The outer layer of DNN mostly uses the softmax module for the solution of most of the classification problems. The softmax formula is also known as normalised exponential, is given below in Eq. (12):

$$Y_i = \frac{exp\,(a_i)}{\sum_j exp\,(a_j)} \tag{12}$$

whereas, j is the set of output nodes, $a_i$ is the net input to a particular output node, and $Y_i$ is the value of output node between range (0, 1).

DNNs models with non-linear behaviour can go up to several abstractions of levels that helps in decision making by transforming original data into higher abstract levels. This process streamlines finding the solution for non-linear and complex functions. Basis of DL is automated learning of features that offer the facility of transfer learning and modularity. Unlike conventional machine learning, training of DL networks requires a large amount of data. Convolutional neural network (CNN) and recurrent neural network (RNN) are the renown deep networks [27, 28].

### 7.1 Convolutional neural network (CNN)

CNN is the popular DL methodology, based on the animal's visual cortex. CNNs are very much similar to ANN that can be observed as the acyclic graph in the form of a well-arranged collection of neurons. Although, in CNNs, the neurons in the hidden layers are only interconnected with a subset of neurons in the preceding layer, unlike regular ANN model. This rare type of interconnectivity enables CNN models to learn the discreet features on an object. CNN models are used for face recognition, scene labelling, image classification, document analysis and many more.

The police department of the Penang Island, Malaysia had installed more than 500 CCTV cameras around the Island and many of them were equipped with face recognition technology, which was developed by IBM. Their main objective was to control crime and capture the wanted criminals [29]. Likewise, in China Pharmaceutical University, to control the student attendance and class discipline the university management installed the facial recognition system in the campus, including the classrooms, labs, library and entrance gates. This overall improved the students' response towards academics [30]. Face recognition technology is based on deep CNN models. This process can be performed by using both supervised and unsupervised approaches but supervised methodologies are mostly preferred. Face recognition is performed by taking an input from video or image and detection is made by taking input to greyscale. The features in greyscale are applied one by one and compared with pixel values. The CNN models give high accuracy than past techniques by overcoming the problems, like light intensity and expressions, with the help of trained models using more training samples [31–33].

### 7.2 Recurrent neural network (RNN)

RNNs are used for the tasks that require consecutive sequential inputs for processing. Initially, training of RNNs was done by using backpropagation. RNNs approach utilises one factor of input, at a time, in sequence by keeping state vector in their hidden nodes, in which implicitly within nodes contains information of all the past value of factors of that sequence. RNNs are dynamic and fairly powerful systems, but during the training process the problem occurs as in gradients of backpropagation algorithm either would shrink or grow at every time step, ultimately they might disappear after many cycles. If we explore RNN, deep feedforward networks will be found having all layers sharing the same weight. RNN lags to the capability of storing information for a long time, and deficiency is known as long-term dependencies. To control this shortcoming, one approach has been introduced with explicit memory known as long short-term memory (LSTM). In this method, particular hidden nodes are used to store the information in the form of input data for a much higher time. LSTM is very much recognised for the better-quality performance in speech recognition systems [1, 27, 28].

Apple's Siri, Amazon's Alexa, Microsoft's Cortana, and Google's Assistant are the most popular voice recognizer tools and they are used for making a phone call, play reminders, alarms, provide driving directions and much more. The speech recognizers are developed on RNN networks, which are based on LSTM-RNN architecture. This gives the RNN models the ability to deal with long-distance patterns and makes them suitable for learning long-span relations. The models are trained end-to-end and output is attained [34, 35]. Other few applications of RNN models are keyphrase recognition, meteorological data updating, speech to text [35–38]. Massachusetts Institute of Technology (MIT) had performed an interesting simulated study on self-driving cars, and its framework was also being developed on the deep reinforced model [39].

## 8. Examples of ANN model using Python

### 8.1 Supervised ANN model

A simple ANN model was developed using Python. The model was designed by using supervised CNN methodology for image classification. Images were collected for training and validation purpose of the model for apples and oranges. For training purpose, 20 images were collected for each (apple and orange), making a total of 40 images. For validation purpose, 10 more images were collected for each, making a total of 20 images. The data for the supervised process, of the ANN model, was arranged in a specific way with a separate folder for each process, i.e., training and validation. In a folder named as 'Training', images of each fruit were placed separately in the folders having their name titles, i.e., 'Apple' and 'Orange', and same was done for 'Validation' folder. In the classification and prediction process, the model output was analysed, for the effectiveness of the results, against two parameters: (1) effect of increasing the number of epochs per run, and (2) the number of hidden layers.

### *8.1.1 Number of epochs per run*

The effect of increasing the number of epochs on the model, for each run, is shown in **Table 2**. The effectiveness of the output is measured against the % accuracy, and % loss for different number epochs. The number of hidden layers for these tests were kept constant for each run.

**Table 2** clearly shows that an increasing number of epochs refines the output by increasing the accuracy and decreasing the data loss. The model gave a correct prediction of the fruit classification in all the runs.

### *8.1.2 Number of hidden layers*

The effect of increasing the number of hidden layers on the model, for each run, is shown in **Table 3**. The effectiveness of the output is measured against the % accuracy, and % loss for various number hidden layers. The number of epochs for these tests was kept constant for each run.

**Table 3** clearly shows that an increasing number of hidden layers increases the model effectiveness by increasing the accuracy and decreasing the data loss. The model gave one wrong prediction, when there were 2 hidden layers. Whereas, by increasing the number of hidden layers, the model started to predict correctly.

### *8.1.3 Overall summary*

The output window from the model is shown in **Figure 13**. It can be seen that the model successfully predicted the correct output ('Apple'). The accuracy of the model was increasing with each epoch from almost 37 to 89% and data loss was also decreasing, consecutively. The program code for this model is given in Appendix A.

| Number of epochs | % Accuracy | % Loss | Prediction |
|---|---|---|---|
| 4 | 74.14 | 56.41 | Correct |
| 8 | 81.25 | 43.44 | Correct |
| 12 | 100 | 27.77 | Correct |

**Table 2.**
*Output summary for increasing number of epochs.*

| Number of hidden layers | % Accuracy | % Loss | Prediction |
|---|---|---|---|
| 2 | 45.83 | 67 | Incorrect |
| 4 | 70 | 64.90 | Correct |
| 6 | 100 | 61.38 | Correct |

**Table 3.**
*Output summary for increasing number of hidden layers.*

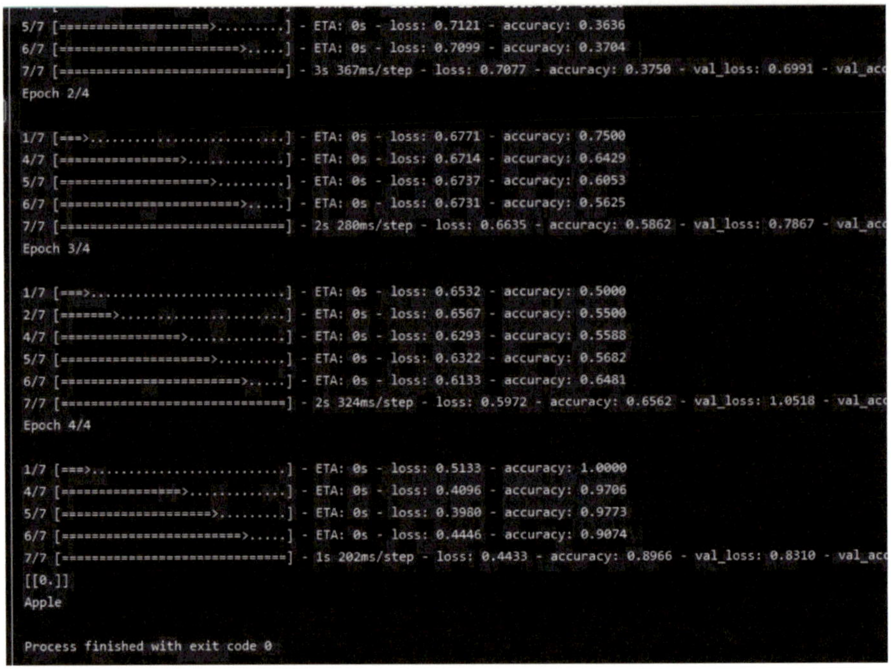

```
5/7 [==================>.........] - ETA: 0s - loss: 0.7121 - accuracy: 0.3636
6/7 [=======================>....] - ETA: 0s - loss: 0.7099 - accuracy: 0.3704
7/7 [============================] - 3s 367ms/step - loss: 0.7077 - accuracy: 0.3750 - val_loss: 0.6991 - val_acc
Epoch 2/4

1/7 [===>........................] - ETA: 0s - loss: 0.6771 - accuracy: 0.7500
4/7 [===============>............] - ETA: 0s - loss: 0.6714 - accuracy: 0.6429
5/7 [==================>.........] - ETA: 0s - loss: 0.6737 - accuracy: 0.6053
6/7 [=======================>....] - ETA: 0s - loss: 0.6731 - accuracy: 0.5625
7/7 [============================] - 2s 280ms/step - loss: 0.6635 - accuracy: 0.5862 - val_loss: 0.7867 - val_acc
Epoch 3/4

1/7 [===>........................] - ETA: 0s - loss: 0.6532 - accuracy: 0.5000
2/7 [=======>....................] - ETA: 0s - loss: 0.6567 - accuracy: 0.5500
4/7 [===============>............] - ETA: 0s - loss: 0.6293 - accuracy: 0.5588
5/7 [==================>.........] - ETA: 0s - loss: 0.6322 - accuracy: 0.5682
6/7 [=======================>....] - ETA: 0s - loss: 0.6133 - accuracy: 0.6481
7/7 [============================] - 2s 324ms/step - loss: 0.5972 - accuracy: 0.6562 - val_loss: 1.0518 - val_acc
Epoch 4/4

1/7 [===>........................] - ETA: 0s - loss: 0.5133 - accuracy: 1.0000
4/7 [===============>............] - ETA: 0s - loss: 0.4096 - accuracy: 0.9706
5/7 [==================>.........] - ETA: 0s - loss: 0.3980 - accuracy: 0.9773
6/7 [=======================>....] - ETA: 0s - loss: 0.4446 - accuracy: 0.9074
7/7 [============================] - 1s 202ms/step - loss: 0.4433 - accuracy: 0.8966 - val_loss: 0.8310 - val_acc
[[0.]]
Apple

Process finished with exit code 0
```

**Figure 13.**
*Output summary of CNN model.*

## 8.2 Unsupervised ANN model

A simple unsupervised ANN model was developed for the colour quantization of an image, using Python, and Self-Organising Maps (SOM) methodology was adopted. SOM is basically used for feature detection.

Two different images of houses were selected for colour quantization by the SOM model. Separate tests were conducted with each image keeping the same model conditions. In each test, the developed SOM model reduced the distinct colours of the image, and another image was developed. This technique helped the model to learn the colours in the image and then use the same colours to reconstruct that image. The pictorial views for each output are shown in **Figure 14**.

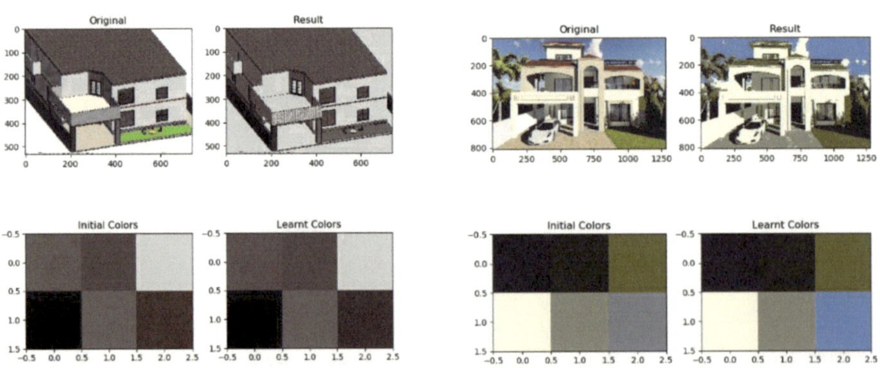


**Figure 14.**
*Pictorial output of SOM model.*

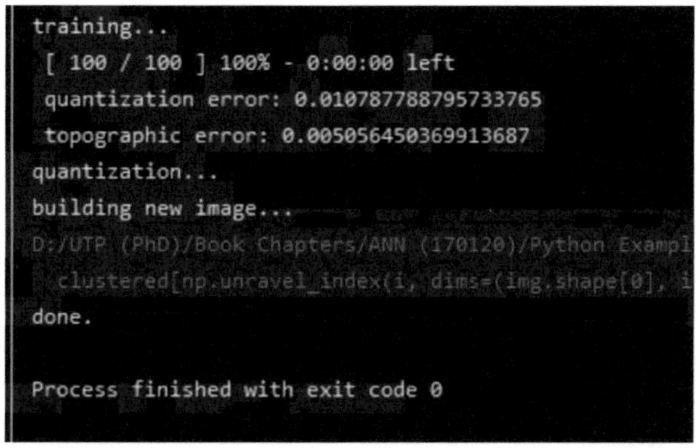


**Figure 15.**
*Output window of SOM model.*

### *8.2.1 Overall summary*

It can be seen in the output results that for each test the model detected the distinct colours and using the same colours it reproduced that image. The output window from the model is shown in **Figure 15**. The program code for this model is given in Appendix B.

## 9. Conclusions

Operation of the ANN model is the simulation of the human brain, and they fall under the knowledge domain of AI. The popularity of ANN models were increased in the early 1990s, and many studies have been done since. The basic ANN model has three main layers, and the main process is performed in the middle layer known as the hidden layer. The output of the ANN model is very much dependent on the characteristics and function it carries under the hidden layer. Among the feedforward and feedback networks, the latter one propagates the error unless it became minimum for more effective results. The ANN models can perform supervised learning as well as unsupervised learning depending upon the task. The DL algorithms are very much popular among researchers because of effective outputs with large data. CNN and RNN are the two renown deep networks, and they have been used for various applications. Output accuracy of the ANN models is very much dependent on the number of hidden layers and the number of epochs.

In this era of automation, the AI plays an important role, and most of the daily use applications are based on the architecture of ANN models. This ANN technology, combined with other advanced and AI knowledge areas, is making life easier in almost every domain. This evolution of DNN models has led to the creation of Sophia the Robot (Hanson Robotics); the journey is on-going.

## Acknowledgements

We will like to acknowledge UTP.

## Conflict of interest

There is no conflict of interest.

## Notes/Thanks/Other declarations

Special thanks to UTP

## Appendix A

Program code for supervised CNN model is given below:

***Step#1*** *Opening Python*

Python was opened, and conda environment was selected.

***Step#2*** *Installing and Import Necessary Data Sources*

```
from keras.preprocessing.image import ImageDataGenerator
from keras.models import Sequential
from keras.layers import Conv2D, MaxPooling2D
from keras.layers import Activation, Dropout, Flatten, Dense
from keras import backend as K
from keras.preprocessing import image
import tensorflow as tf
import numpy as np
```

***Step#3*** *CNN Convolutional Network Model*

```
#Input Layer
model = Sequential()

# Convolutional Layer 1
model.add(Conv2D(64 (3, 3), input_shape = (150, 150, 3), activation = 'relu'))
model.add(MaxPooling2D(pool_size = (2, 2)))

# Convolutional Layer 2
model.add(Conv2D(64, (3, 3), activation = 'relu'))
model.add(MaxPooling2D(pool_size = (2, 2)))
```

The convolutional network helps to extract features from the image and digit 64 means to extract 64 features.

```
model.add(Flatten())

# Hidden Layer 1
model.add(Dense(units = 64, activation = 'relu'))

# Hidden Layer 2
model.add(Dense(units = 32, activation = 'relu'))
```

64 and 32 represents the number of neurons in these layers.

```
# Output Layer
model.add(Dense(units = 1, activation = 'sigmoid'))

# Compiling the CNN
model.compile(optimizer = 'adam', loss = 'binary_crossentropy', metrics = ['accuracy'])
```

Sigmoid represents the activation function of this model.

***Step#4** Fitting CNN to the Images (Training and Validation)*

```
train_datagen = ImageDataGenerator(rescale = 1./255,
shear_range = 0.2,
zoom_range = 0.2,
horizontal_flip = True)
test_datagen = ImageDataGenerator(rescale = 1./255)
training_set = train_datagen.flow_from_directory('Data/Train',

target_size = (150, 150),
batch_size = 12,
class_mode = 'binary')
test_set = test_datagen.flow_from_directory('Data/Validation',

target_size = (150, 150),
batch_size = 8,
class_mode = 'binary')
model.fit_generator(training_set,
steps_per_epoch = 10,
epochs = 4,
validation_data = test_set,
validation_steps = 10)
```

directories are for the path to the training folder and validation folder

***Step#5** Running the ANN Model*

```
test_image = image.load_img('Data/Apple.JPG', target_size = (150, 150))
test_image = image.img_to_array(test_image)
test_image = np.expand_dims(test_image, axis = 0)
result = model.predict(test_image)
training_set.class_indices

if result[0][0] == 1:
  prediction = 'Orange'
else:
  prediction = 'Apple'

print (prediction)
```

# Appendix B

Program code for unsupervised SOM model is given below:

***Step#1** Opening Python*

Python was opened, and conda environment was selected.

***Step#2** Installing and Import Necessary Data Sources*

```
from minisom import MiniSom
import numpy as np
import matplotlib.pyplot as plt
```

***Step#3** Importing Image*

```
img = plt.imread('HouseTest2.jpg')
```

Image path is to be given here.

***Step#4** SOM Model*

```
# Reshaping the pixels matrix
pixels = np.reshape(img, (img.shape[0]*img.shape[1], 3)) / 255.

# SOM initialization
som = MiniSom(2, 3, 3, sigma=1.,
learning_rate=0.2, neighborhood_function='bubble')

# Setting Weights
som.random_weights_init(pixels)
starting_weights = som.get_weights().copy()
som.train_random(pixels, 100, verbose=True)
```

100 is the number of training iteration

```
# Quantization
qnt = som.quantization(pixels)

# Compilation
clustered = np.zeros(img.shape)

for i, q in enumerate(qnt):
  clustered[np.unravel_index(i, dims=(img.shape[0], img.shape[1]))] = q
print('done.')
```

***Step#5** Running and Plotting of ANN Model*

```
plt.figure(figsize=(7, 7))
plt.subplot(221)
plt.title('Original')
plt.imshow(img)
plt.subplot(222)
plt.title('Result')
plt.imshow(clustered)

plt.subplot(223)
plt.title('Initial Colors')
plt.imshow(starting_weights)
plt.subplot(224)
plt.title('Learnt Colors')
plt.imshow(som.get_weights())

plt.tight_layout()
plt.show()
```

*DOI: http://dx.doi.org/10.5772/intechopen.91935*

## Author details

Wesam Salah Alaloul* and Abdul Hannan Qureshi
Department of Civil and Environmental Engineering, Universiti Teknologi PETRONAS, Perak, Malaysia

*Address all correspondence to: wesam.alaloul@utp.edu.my

## References

[1] Ciaburro G, Venkateswaran B. Neural Networks with R: Smart Models Using CNN, RNN, Deep Learning, and Artificial Intelligence Principles. Birmingham: Packt Publishing; 2017

[2] Zou J, Han Y, So S-S. Overview of Artificial Neural Networks. Vol. 458. Totowa: Humana Press; 2008. pp. 14-22. DOI: 10.1007/978-1-60327-101-1_2

[3] Aggarwal CC. Neural Networks and Deep Learning. Cham: Springer International Publishing; 2018. DOI: 10.1007/978-3-319-94463-0

[4] Araghinejad S. Data-Driven Modeling: Using MATLAB® in Water Resources and Environmental Engineering. Vol. 67. Dordrecht: Springer Netherlands; 2014. DOI: 10.1007/978-94-007-7506-0

[5] Adeli H, Yeh C. Perceptron learning in engineering design. Computer-Aided Civil and Infrastructure Engineering. 1989;**4**:247-256

[6] Adeli H. Neural networks in civil engineering: 1989–2000. Computer-Aided Civil and Infrastructure Engineering. 2001;**16**:126-142. DOI: 10.1111/0885-9507.00219

[7] Mehrotra K, Mohan CK, Ranka S. Elements of Artificial Neural Networks. Cambridge: MIT Press; 1997

[8] Rumelhart DE, Hinton GE, Williams RJ. Learning internal representations by error propagation. California Univ San Diego La Jolla Inst for Cognitive Science; 1985

[9] Schmidhuber J. Deep learning in neural networks: An overview. Neural Networks. 2015;**61**:85-117. DOI: 10.1016/j.neunet.2014.09.003

[10] Hochreiter S, Schmidhuber J. Long short-term memory. Neural Computation. 1997;**9**:1735-1780. DOI: 10.1162/neco.1997.9.8.1735

[11] Feldman JA, Fanty MA, Goddard NH. Computing with structured neural networks. IEEE Computer. 1988;**21**:91-103

[12] Profillidis VA, Botzoris GN. Artificial intelligence—Neural network methods. Modeling of Transport Demand. Elsevier; 2019:353-382. DOI: 10.1016/B978-0-12-811513-8.00008-X

[13] Taylor M. Make Your Own Neural Network: An In-Depth Visual Introduction for Beginners. 2017. Independently published

[14] Priddy KL, Keller PE. Artificial Neural Networks: An Introduction. Bellingham: SPIE Press; 2005

[15] Suk H-I. An Introduction to Neural Networks and Deep Learning. 1st ed. Cambridge: Academic Press; 2017. DOI: 10.1016/B978-0-12-810408-8.00002-X

[16] Poznyak TI, Chairez Oria I, Poznyak AS. Background on dynamic neural networks. Ozonation and Biodegradation in Environmental Engineering. 2019:57-74. DOI: 10.1016/b978-0-12-812847-3.00012-3

[17] Zamir AR, Wu TL, Sun L, Shen WB, Shi BE, Malik J, et al. Feedback networks. In: Proceedings of the 30th IEEE Conference on Computer Vision and Pattern Recognition (CVPR 2017); 2017. pp. 1808-1817. DOI: 10.1109/CVPR.2017.196

[18] Whittington JCR, Bogacz R. Theories of error back-propagation in the brain. Trends in Cognitive Sciences. 2019;**23**: 235-250. DOI: 10.1016/j.tics.2018.12.005

[19] Awodele O, Jegede O. Neural networks and its application in

engineering. In: Proceedings of the 2009 InSITE Conference; 2009. DOI: 10.28945/3317

[20] Rao Z, Alvarruiz F. Use of an artificial neural network to capture the domain knowledge of a conventional hydraulic simulation model. Journal of Hydroinformatics. 2007;**9**:15-24. DOI: 10.2166/hydro.2006.014

[21] Perez RR, Marques A, Mohammadi F. The application of supervised learning through feed-forward neural networks for ECG signal classification. In: Canadian Conference on Electrical and Computer Engineering; 2016. pp. 1-4. DOI: 10.1109/CCECE.2016.7726762

[22] Daniel G. Principles of Artificial Neural Networks. Singapore: World Scientific; 2013

[23] Gurney K. An Introduction to Neural Networks. Boca Raton: CRC Press; 1997

[24] Damilola S. A review of unsupervised artificial neural networks with applications. International Journal of Computers and Applications. 2019; **181**:22-26. DOI: 10.5120/ijca2019918425

[25] Coulibaly P, Anctil F, Bobée B. Daily reservoir inflow forecasting using artificial neural networks with stopped training approach. Journal of Hydrology. 2000;**230**:244-257

[26] Liang P, Bose NK. Neural Network Fundamentals with Graphs, Algorithms and Applications. New York: MacGraw-Hill; 1996

[27] Chauhan NK, Singh K. A review on conventional machine learning vs deep learning. In: 2018 International Conference on Computing, Power and Communication Technologies (GUCON 2018); 2019. pp. 347-352. DOI: 10.1109/GUCON.2018.8675097

[28] Tavanaei A, Ghodrati M, Kheradpisheh SR, Masquelier T, Maida A. Deep learning in spiking neural networks. Neural Networks. 2019;**111**:47-63. DOI: 10.1016/j.neunet.2018.12.002

[29] This Malaysian State is Using Facial Recognition to Catch Criminals—Tech. Available from: https://sea.mashable.com/tech/1725/this-malaysian-state-is-using-facial-recognition-to-catch-criminals [Accessed: 27 February 2020]

[30] Facial Recognition System in University Classrooms Sparks Controversy—SHINE News. Available from: https://www.shine.cn/news/nation/1909041394/ [Accessed: 27 February 2020]

[31] Karahan Ş, Yildirim MK, Kirtaç K, Rende FŞ, Bütün G, Ekenel HK. How image degradations affect deep CNN-based face recognition? 2016 Int. Conf. Biometrics Spec. Interes. Gr., Vol. P-260. IEEE; 2016. pp. 1-5. DOI: 10.1109/BIOSIG.2016.7736924

[32] Khan S, Javed MH, Ahmed E, Shah SAA, Ali SU. Networks and implementation on smart glasses. In: 2019 International Conference on Information Science and Communication Technologies; 2019. pp. 1-6. DOI: 10.1109/CISCT.2019.8777442

[33] Bhandare A, Bhide M, Gokhale P, Chandavarkar R. Applications of convolutional neural networks. International Journal of Computer Science and Information Technologies. 2016;7:2206-2215

[34] Zhang D, Wang D. Relation classification: CNN or RNN? Lecture Notes in Computer Science (including subseries Lecture Notes in Artificial Intelligence and Lecture Notes in Bioinformatics). 2016;**10102**:665-675. DOI: 10.1007/978-3-319-50496-4_60

[35] Graves A, Mohamed AR, Hinton G. Speech recognition with deep recurrent